T0322933

SACRED GEOMETRY
OF THE
STARCUT DIAGRAM

SACRED GEOMETRY OF THE STARCUT DIAGRAM

THE GENESIS OF NUMBER, PROPORTION, AND COSMOLOGY

MALCOLM STEWART

Inner Traditions
Rochester, Vermont

Inner Traditions
One Park Street
Rochester, Vermont 05767
www.InnerTraditions.com

Originally published in the United Kingdom in 2009 by Floris under the title *Patterns of Eternity: Sacred Geometry and the Starcut Diagram*
First U.S. edition published in 2022 by Inner Traditions

Cataloging-in-Publication Data for this title is available from the Library of Congress

ISBN 978-1-64411-430-8 (print)
ISBN 978-1-64411-431-5 (ebook)

Printed and bound in India by Replika Press, Pvt. Ltd.

10 9 8 7 6 5 4 3 2 1

Text design and layout by Debbie Glogover
This book was typeset in Garamond Premier Pro with Gill Sans MT Pro and Hermann used for display typefaces

To send correspondence to the author of this book, mail a first-class letter to the author c/o Inner Traditions • Bear & Company, One Park Street, Rochester, VT 05767, and we will forward the communication.

This book is dedicated to my children, Adam, Mark, Martha, Rosie, and Phil.

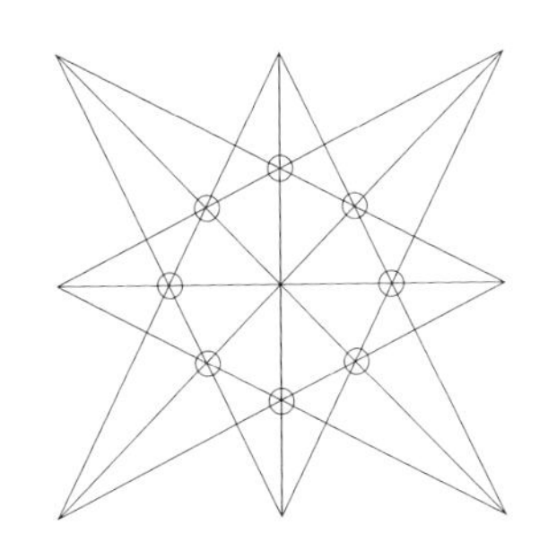

Contents

Acknowledgments

I have acknowledged people to whom I am indebted at the relevant points in the text, and there is a full bibliography at the end. But at the outset I particularly want to say thanks to the following: the late Sivakalki Swami, who drew my attention to the Vedic name Agni; the late John Michell, for his insight into geomancy and traditional measures; Keith Critchlow, from whom I learned the basics of sacred geometry; Oscar Ichazo, for integrating the ancient principles of harmony, number, and form in a contemporary school of philosophy.

Thanks also to my friend Elliott Manley, skeptic, mathematician, hawkeyed editor, and sounding board. Thanks to Allan Brown, designer and author, without whose supportive interest this book may never have been written. Thanks to Rob Leech who helped me run the last mile.

Thanks to my editor, Christopher Moore, and Sally Martin and Helena Waldron of Floris Books; their creative input has greatly enhanced the book's beauty.

A very personal thank-you goes to my wife Nora whose patience and trust made it possible to pursue this study of many years, and whose feedback has been a sure intuitive guide. She, together with friends Barbara, Natalie, Celia, Brian, and Patrick, has been a trusted lifeline in more ways than one.

Thank you all.

ILLUSTRATIONS CREDITS

I am grateful to the following for the use of images in this book. Apart from those below there are a number that were found, unattributed, on the internet. If I have inadvertently omitted an acknowledgment, I apologize and will rectify the omission in the event of a reprinting.

Thanks to Allan Brown for the photo used in the prologue and epilogue; for figures 1.2, 1.9, 2.8, 4.8, 7.2, 19.1, 19.2, 20.4, 25.12, and 25.16.

Thanks to Barbara Carlisle for the photo in figure 1.14.

Thanks to Oscar Ichazo and the Arica Institute for permission to reproduce figure 5.17.

Thanks to John Martineau for figure 24.2.

Thanks to Ashen Venema for the figure 24.5 illustration.

For figure 25.1, thanks to qian jin/iStockphoto.

The following images were found on the internet under a Creative Commons or GNU license: figure 1.1, Luc Viatour; figure 3.1, David Castor; figure 4.1, Ondrej Žvácek; figure 6.1, Ricardo Liberato; figure 9.1, Thomas Wydra; figure 11.1, Slyvie Mayrargue; figure 12.1, Oni Lukos; figure 13.1, Nick Hobgood; figure 15.1, asqueraid; figure 16.1, NASA and ECA; figure 18.1, Floris Books; figure 21.1, Patrick Rouzet; figure 23.1, Alastair Moore.

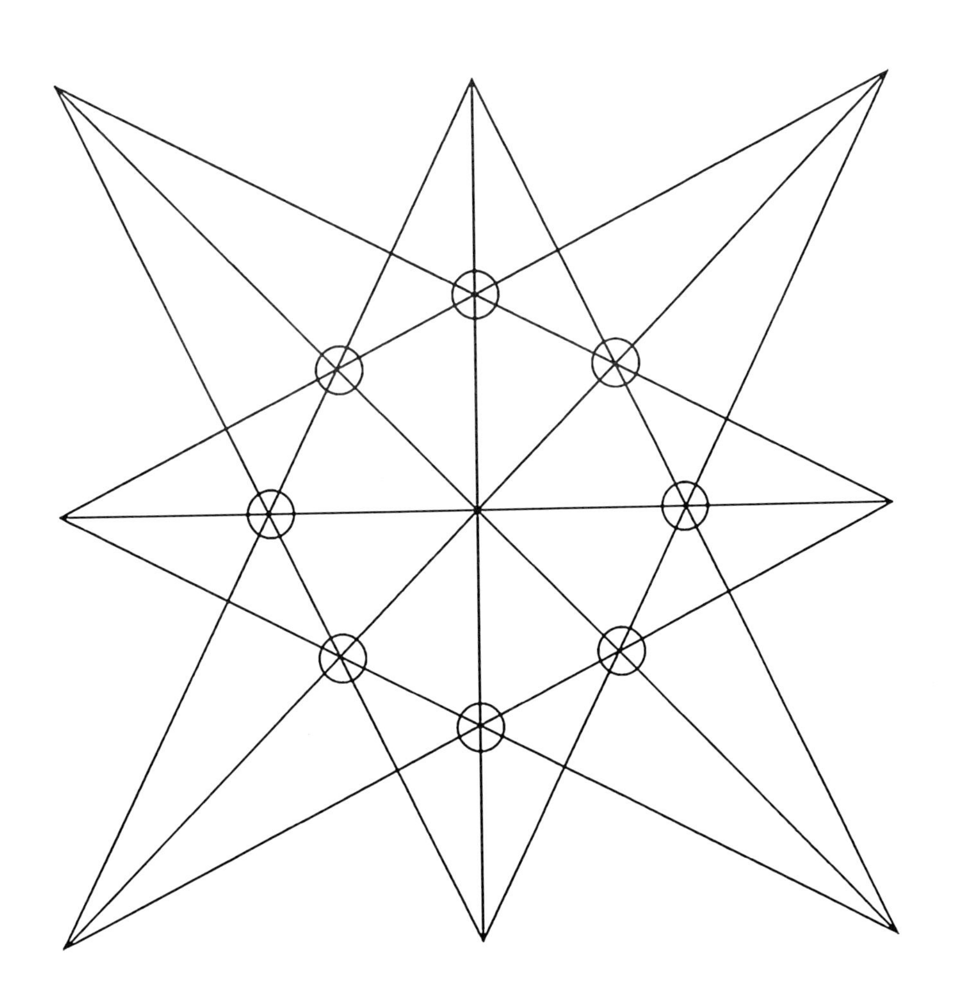

Sandwood Bay, Cape Wrath, Scotland.

Prologue

The two men walked along a ridge following the sweep of the beach around the bay. Ahead of them the westering sun caught the headlands south of Cape Wrath. As they took in the lines of the cliffs and the timeless ocean horizon, their talk had moved on from the earthbound beauty of their surroundings to the lofty arts of number, harmony, and form. The monk had explained how the ancients had believed such arts readied the soul for higher things, bringing it to the very shoreline of the eternal. He had described the strange and secret ways in which, he said, the old priests and philosophers had taught their pupils.

"So you're telling me," said his burly southern friend, "that all that stuff about right angles and hypotenuses and the musical theory about lengths of strings and golden measures, sacred numbers, and heaven knows what else—all the stuff we think of as coming from Pythagoras—that the lot was taught to him through just one simple diagram, when he was in Egypt?"

"It's certainly possible," said the monk, "and much else besides—surveying, architectural ground plans, cosmology, theology—even their very system of numbers itself. I think there was a geometric core to much that was taught in the 'House of Life' in Heliopolis. And of course ancient Chaldea and India had it, too, and China no doubt."

The southerner looked at the Benedictine's windbeaten face. Was he joking? One simple diagram?

"Next you'll be telling me that it was used to build the Great Pyramid."

"Well, not 'build' it exactly, but certainly for planning the ground-square and later, to get the height and the face angles for cutting into the stone."

"I thought that was some obscure angle—just under fifty-two degrees . . ." The layman was not about to be convinced.

"Yes, that's about it," said the monk cheerfully. "It's right there in the diagram; which also divides itself into sevens and elevens and thirteens and an infinity of other sections. If Pythagoras had had the good fortune to be a Scotsman he could have used it for designing tartans! And if Hermann Hesse had only known about it he might have had a real glass bead game to play. But you're wrong about it being complicated. It's so easy that any half-skilled journeyman could sketch it out as a template with no problem at all."

The southerner's silence proclaimed his skepticism.

"Here," said the monk. "I'll show you."

He clambered down an incline from the grass ridge to the banked sea gravel below. Slipping his sandals off, he walked out onto the beach. The receding tide had left the surface smooth and damp. There he squatted with his cowl flapping in the breeze. He reached for a twig of driftwood and in just a few strokes, he drew out a rudimentary boxed star-shape in the sand. His friend stood waiting for more, but the monk, it seemed, had finished.

"So . . . is that it?"

"Yes, that's it," said the monk.

"Just that simple?"

"Yes—just that simple."

Opposite:
The Starcut diagram.

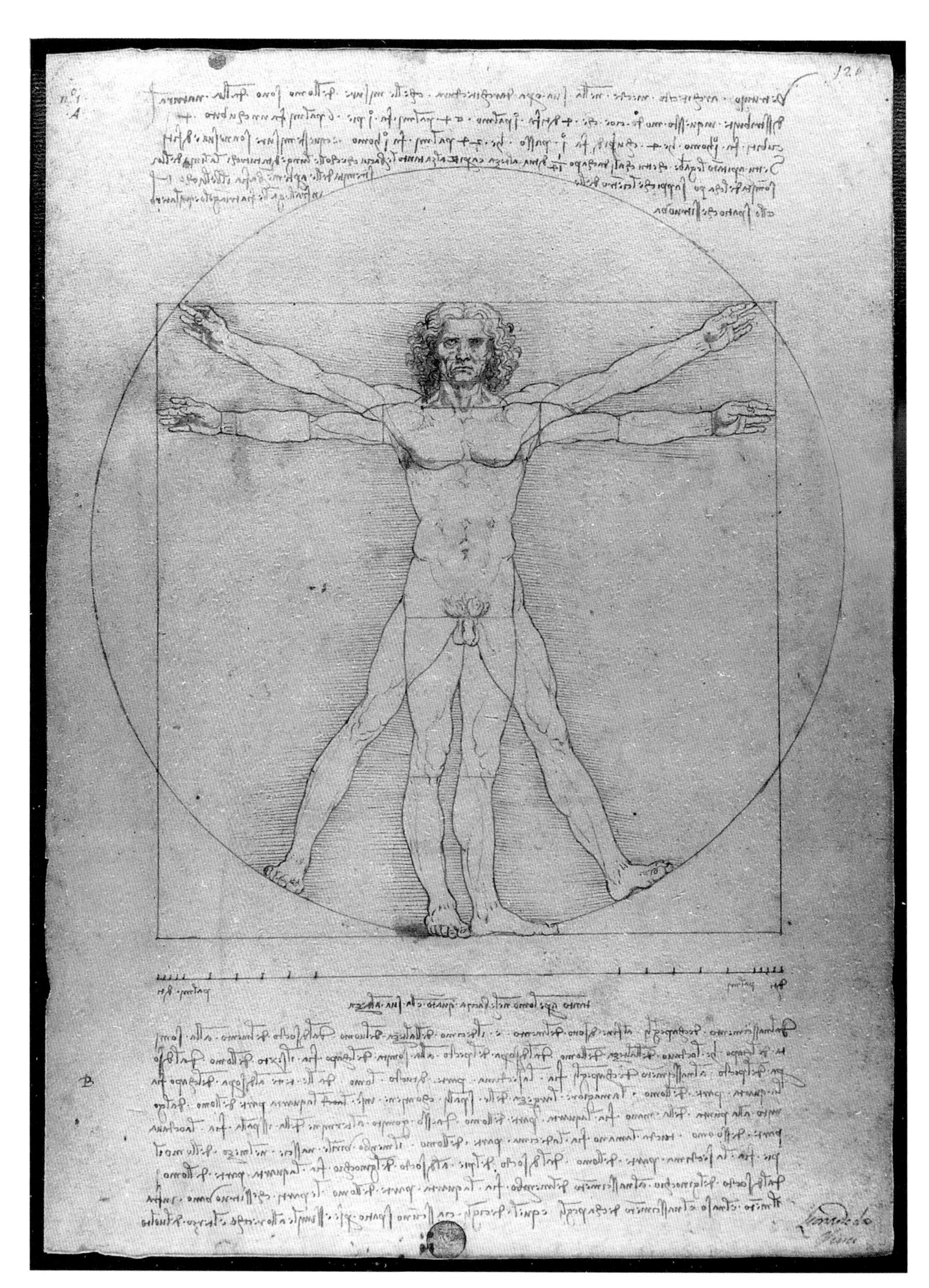

Figure 1.1

Vitruvian Man, Leonardo da Vinci.

1

An Introduction to Sacred Geometry

Who Tastes Knows

Thomas Gradgrind, sir. A man of realities. A man of facts and calculations. A man who proceeds upon the principle that two and two are four, and nothing over, and who is not to be talked into allowing for anything over. Thomas Gradgrind, sir . . . with a rule and a pair of scales . . . ready to weigh and measure. . . . It is a mere question of figures, a case of simple arithmetic.

Charles Dickens,
Hard Times

Pythagoras is said to have opened his teaching by cutting an apple across the core to show the pentagonal star arrangement of its seeds, thus illustrating the difference between number as sheer quantity and number as quality. A group of five apples in a dish illustrates accidental quantity—it could equally be four, seven, or however many—whereas a closer look shows that five has a more intimate or "qualitative" relationship to apples.

Figure 1.2
Pythagoras's apple.

Nowadays there is a lot of published data on a topic called "sacred geometry." Much of it rehashes what is already well known or wanders into distant esoteric borderlands where few can follow. I know from giving presentations on the subject that while a growing number of people buy books on it, very few possess, let alone use, a square and compasses (or indeed the common modern substitute—a computer graphics program). That's a pity because actually exploring number, proportion, and form by drawing, in the ancient manner, is a most enjoyable and effective way of discovery, always surpassing the accumulation of the data of others.

Our reluctance to draw may have arisen from the very first geometry lesson that was taught at school—the "perpendicular bisector." What a mean little device to use as an introduction to one of the world's most gloriously beautiful studies!

The minimal approach in figure 1.3 has unjustly become associated with Euclid who laid out geometry in proven logic, from the basic axioms arising from the very first point on a page right through to the complex volumetric properties of a twelve-faced dodecahedron. For such a task it is necessary to isolate every detail, and a geometric shorthand helps toward this. In fact the first problem that Euclid set himself was the construction of a triangle with sides of equal length. Hence, at the outset of his work, he used the full vesica piscis, not the spindly construction that has been favored in thousands of schoolrooms since his time (figs. 1.4, 1.5).

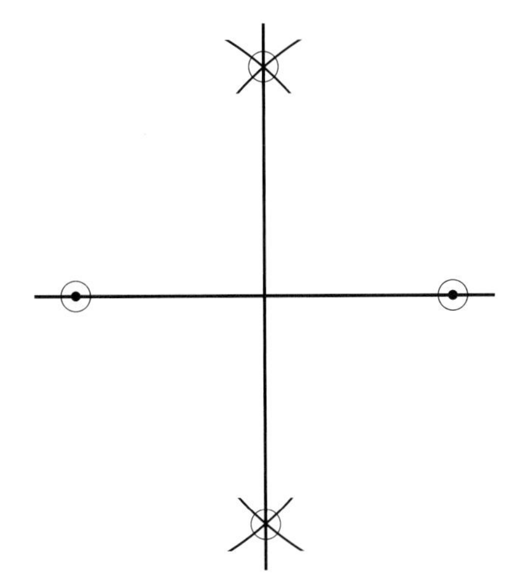

Figure 1.3
Here is the spindly perpendicular bisector we learned at school. It comes from the beautiful vesica piscis (Latin for "the bladder of a fish") shown in figure 1.4.

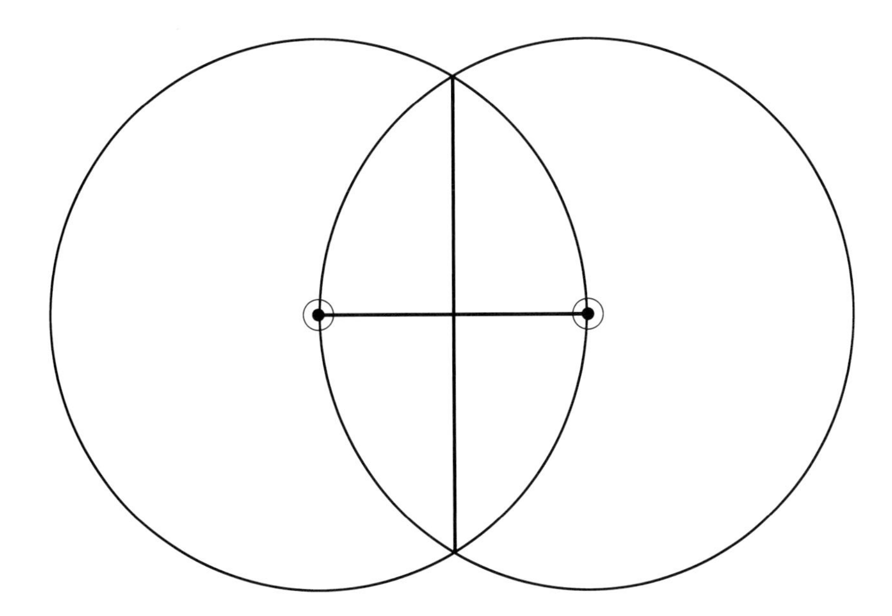

Figure 1.4
The vesica is composed of similar circles whose circumferences pass through each other's centers.

The human body itself, in its motion as much as its anatomical proportions, is archetypally geometric. The sockets of our skeletal joints make circular movements and arcs that are easy and natural for us.

As an experiment, stand in front of a wall with a piece of chalk in each hand. Now, by simply letting both arms circle from their shoulder sockets while taking a steady breath in and out, and letting the chalk trace the movement, you will find that your own body's version of the vesica piscis will appear. For this experiment it is best to close your eyes and simply trust the form of your body. You may be amazed by the accuracy of the circles you make.

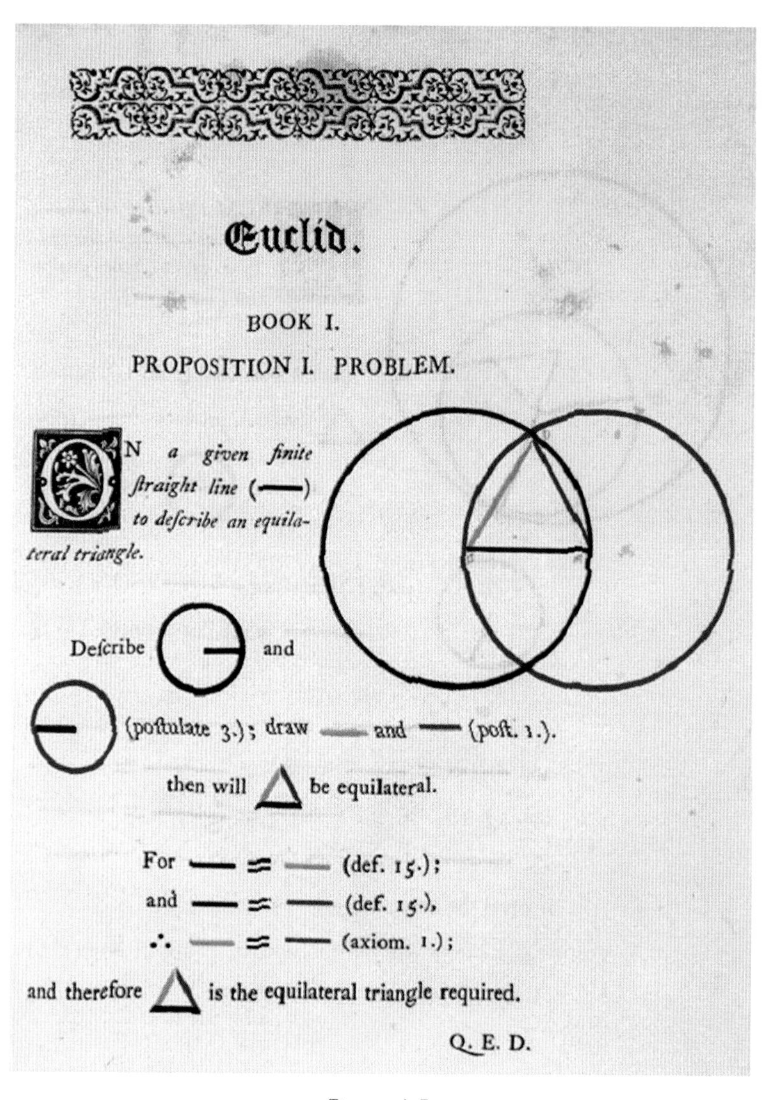

Euclid.

BOOK I.

PROPOSITION I. PROBLEM.

ON *a given finite ſtraight line* (—) *to deſcribe an equilateral triangle.*

Deſcribe and (poſtulate 3.); draw — and — (poſt. 1.).

then will △ be equilateral.

For — = — (def. 15.);
and — = — (def. 15.),
∴ — = — (axiom. 1.);

and therefore △ is the equilateral triangle required.

Q. E. D.

Figure 1.5

This page of Euclid is from Oliver Byrne's edition of 1847, where colors were used to code and illuminate the diagrams.

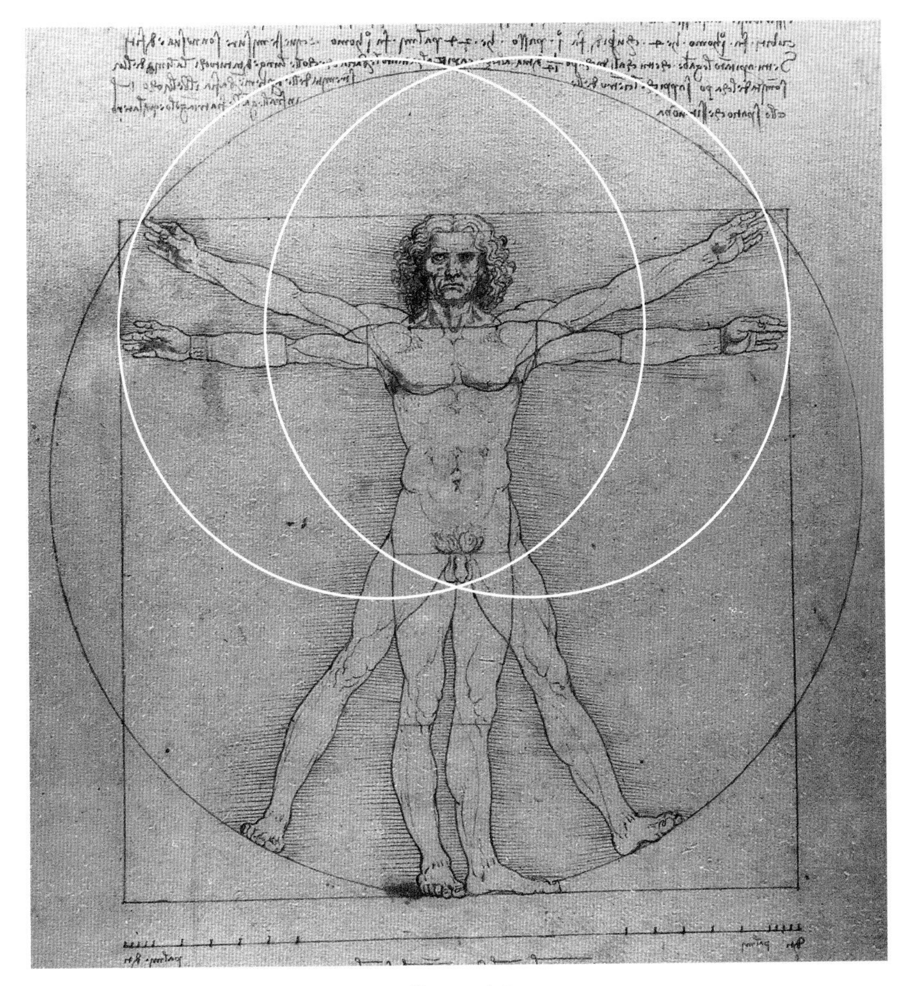

Figure 1.6

If the Vitruvian Man is allowed to complete the movement that Leonardo's drawing suggests him to have started, he will not make the single circle of the drawing. He will make a vesica with its circles' centers at his shoulder sockets and its radii as the length of his arms. Since the width of the shoulders is not the same as the length of the arms, this is a variant of the classic vesica form.

Traditional systems of dance, movement, and posture, such as ballet, tai chi, and yoga, use circles, arcs, twists, spirals, and swivels. A modern exercise method, Oscar Ichazo's Psychocalisthenics, which works the entire musculoskeletal system, comprises twenty-three movements that all employ spirals, arcs, and circles.

I think the basic appreciation of geometry, in most people, has been obstructed, rather than enhanced, by the overly technical and disembodied way it was first taught to us. School geometry too quickly got bogged down in analytical proofs and theorems rather than illustrating the beautiful order that lies on the interface between mind and structure. Logic is all very well, but people are interested in the logic of forms that they find beautiful. First and foremost I came to this subject as an artist who loves pattern.

Beauty is no optional extra. In the Platonic sense it is an essential secret of the study. What exactly is it that so many people find beautiful in geometry? Socrates was said to have been set on his course to wisdom by being told to contemplate the essential nature of beauty. There is clarity, universality, precision, and a deep "familiarity" between the forms of geometry and the mind's own most reasonable forms of thought. Beauty is difficult to define, and it seems pretentious to try. However, the beauty of fundamental forms and harmonies is always there, speaking directly to a perception that transcends passing fashion and subjective preference. Plato said that the soul was purified by contemplating the mathematical disciplines. For me, the simple beauty of circles themselves is the first evidence of this.

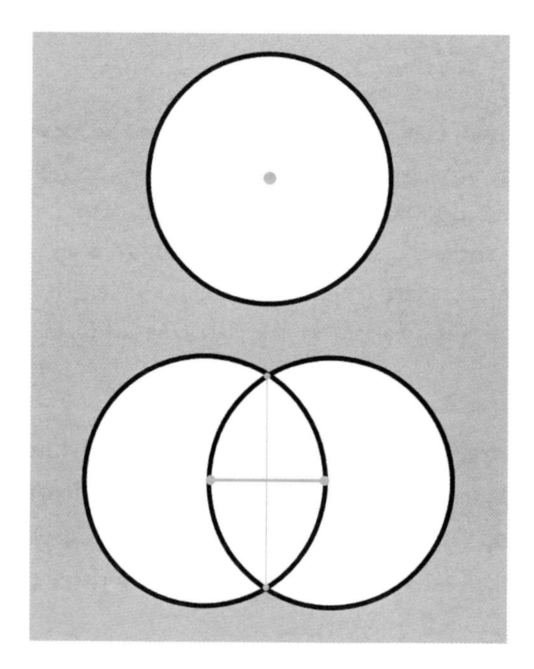

Figure 1.7
Simple beauty.

"Sacred geometry" was a term used by Vitruvius, the Roman architect, to describe the design art used in the building of temples. Over time its meaning has extended, and it has become a nickname for much more. I recall Professor Keith Critchlow saying that if one's view of the cosmos is sacred then the measurement of that cosmos is a sacred activity. Perhaps that view is not shared by so many in the West today, but such a design art makes available a whole range of metaphors that are helpful in grasping some deep ideas. Also this kind of geometry is very much a living art (just look at the crop circles!), and there is much still to be discovered and much beauty to be found.

Discovering things for ourselves helps us to remember them. Nowadays memory is propped up, and seemingly rendered almost unnecessary, by electronic storage devices, whereas the art of memory is said to have been the most ancient of all esoteric skills, that of the bards. In Hellenic myth, Zeus, the animating consciousness, and Mnemosyne, the faculty of memory, were the parents of the nine creative Muses. Early sources carry accounts of heroic feats of memory. Poetic contests required that the bards knew entire epics. The devices of poetry, rhyme, assonance, rhythm, and repetition were ways of assisting in this oral tradition. From very ancient times it has been known that memory works most effectively when the material to be remembered is linked within a context. The context could be one of a narrative, where the material emerges naturally from the story, or it might be a visualized scene—the *loci* ("placing") method in which elements to be remembered are located within the image of a familiar place. Such methods are analogous to what is now known about the way the brain develops neuronal linkages. Memory expert Tony Buzan's widely used "mind-mapping" methods are based on just that. So it is that geometric art with its beauty and its linked logic is a tremendous aid for recalling the lore associated with number, measure, and symbolism.

In the minimal construction of the perpendicular bisector that was taught in school this art was already lost; we were already boxed in by the Gradgrind mentality—a bare-minimum factuality that, being narrowly focused, hid the bursting richness of information revealed in the complete diagram of the vesica piscis, a form that all of us know from

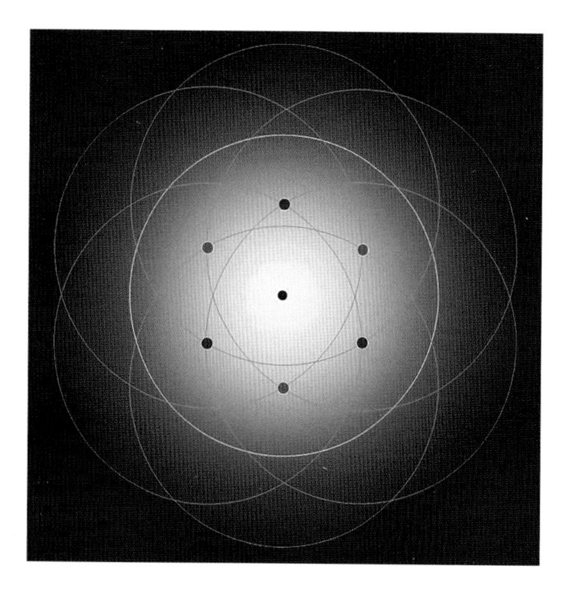

Figure 1.8
The third multiplication of the human cell.

our very cells, whose spheres divided in this selfsame way when we first took shape in the womb as a fertilized egg. Here we see a stylized view of the cell at its third multiplication (fig. 1.8).

The symbol of Saint Edmund of Abingdon, an archbishop of Canterbury in the 1300s, was a triple vesica. It has the strange quality when pictured, as here in figure 1.9, in that while all three circles are interlocked, no two circles are connected. So if one takes any circle away, the other two fall apart.

Figure 1.9
Saint Edmund's rings.

Figure 1.10
Some variations on clusters of circles used in late medieval architecture. Here, alternate points of the dotted-line stars mark the centers of the circles, which touch rather than interlock.

I spent twenty-one years under Edmund's aegis. He happened to be the patron saint of my junior school, secondary school, and the theological college where I spent six years in studies for the priesthood. He was also patron of the hall of residence, now a college, where I lived while at Cambridge University. Edmund was a leading mathematician/geometer in his time. He studied and taught in Paris and Oxford. The triple vesica, it is said, recalled the vision of a triple sun that he had seen as a child. Certainly this form, with variations, was characteristic of the trefoil mullions of much Gothic architecture.

Saint Edmund's rings were my first introduction to a form of geometry that was conceptual and metaphorical at the same time. As noted, no two rings are connected—take any one away and the other two fall apart; but together they make a beautiful interlocked unity. In the Christian faith they symbolize the Trinity. In fact, Edmund's rings involved the same geometry that was taught by our mathematics teacher as Euclid's perpendicular bisection of a line. None of us noticed. And, inevitably, our classics teacher never mentioned that the same vesica form, deployed six times around its original circle, underlay the Tetraktys (see fig. 1.12), the key teaching device of Pythagoras; nor that it is the basic grid for many Roman and Arabian mosaics.

A similar deployment of circles underlies the circle cluster recently popularized as the "Flower of Life" (fig. 1.11). This is a form that has been inscribed haphazardly, seemingly as a piece of Hellenic graffiti, with faint and unclear early Greek writing alongside it, on a granite column of the Osirion, an ancient structure abutting the later temple of Seti at Abydos in Egypt (fig. 1.13).

Figure 1.11
The "Flower of Life."

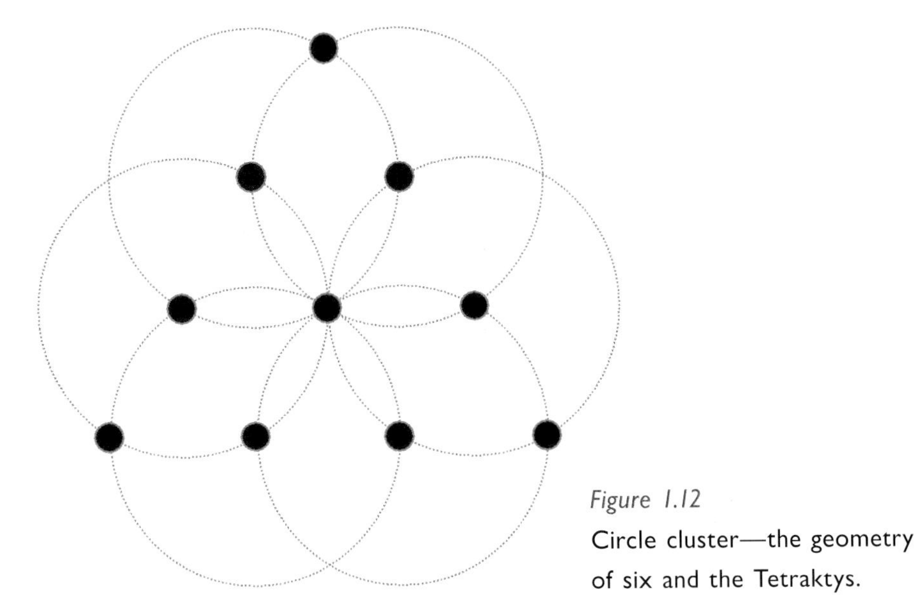

Figure 1.12
Circle cluster—the geometry of six and the Tetraktys.

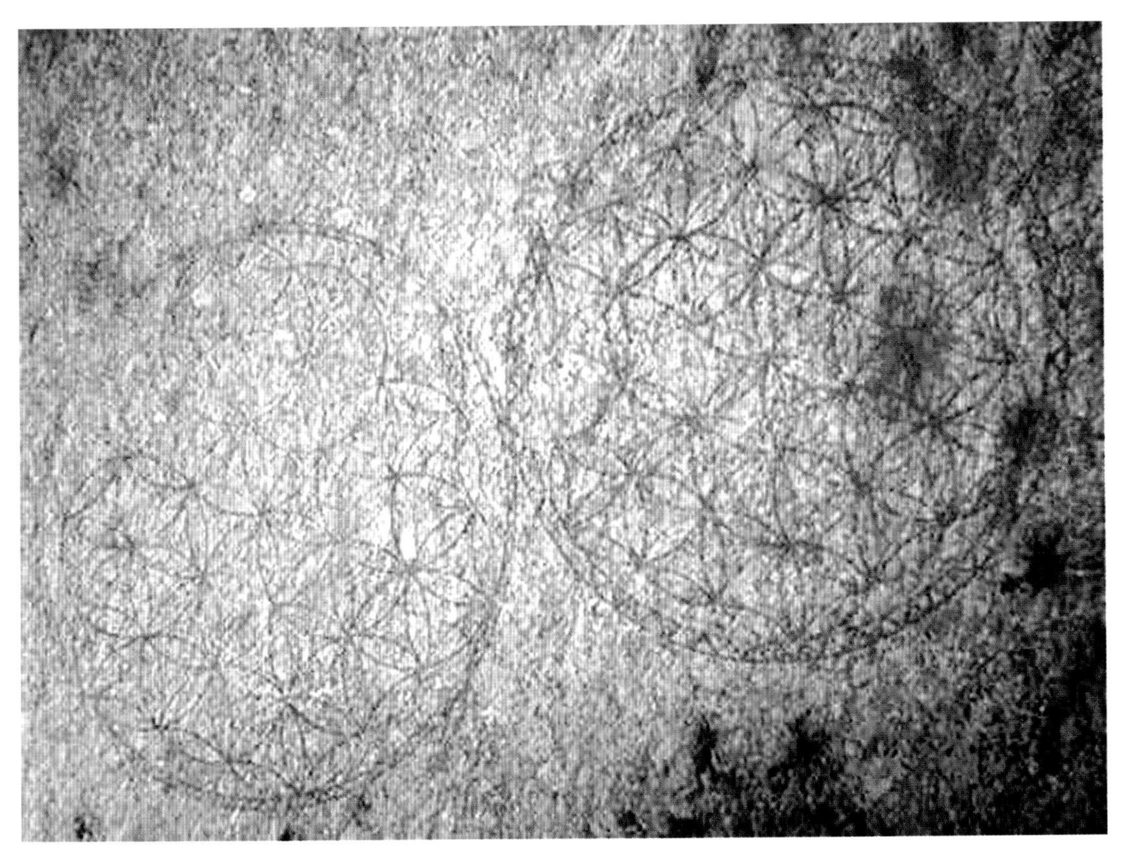

Figure 1.13
Graffiti on the masonry of the Osirion.

Figure 1.14
The relief pavement from Nineveh, now in the Louvre, Paris.

The further example was part of the doorway pavement into the great hall of the palace of Ashurbanipal in Nineveh (fig. 1.14). With lilies and daisies as its surrounds, it suggests water. Those crossing this threshold would thus have their feet ceremonially cleansed as if by walking through a shallow pool.

One of the delights of this whole subject is precisely the way similar figures turn up in many different guises and find application to other fundamentals. For instance, the lacework of a triple vesica becomes the seven zones of radiant and reflected light (fig. 1.15). This version of the figure never struck me until, many years after school and college when, being engaged in stained glass design, I started to think about the difference between combinations of radiant light passing through window glass and those of reflected light from painted surfaces. As clear lights, the three primary colors make white; as pigments, they make black.

It is often in conjunction with some practical application of the principles of form, measure, and vibration that one's theoretical understanding is expanded. The early development of academic geometry is a case in point. From all the indications geometry, and geometrically guided arithmetic, had been a thriving cultural interest for at least a couple of thousand years before the time of Pythagoras in the sixth century BCE. The building of the ziggurats of Mesopotamia, and the pyramids and great temples of Egypt, happened throughout two and more millennia

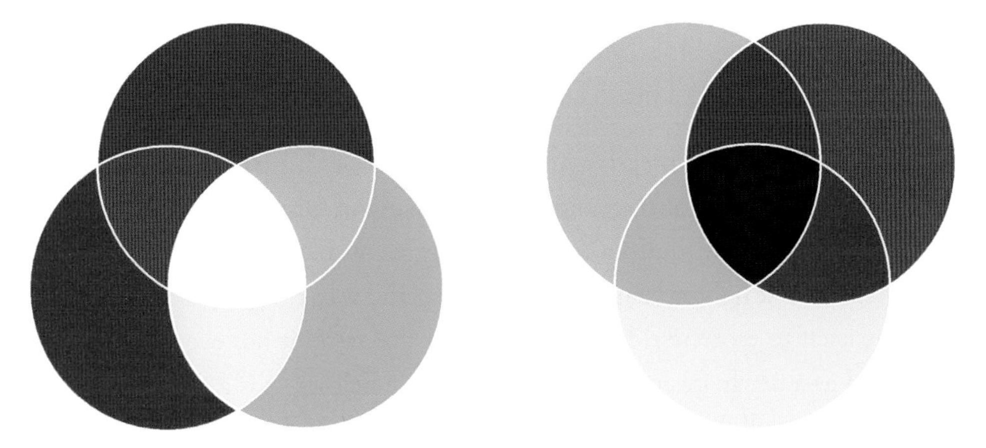

Figure 1.15
Combinations of radiant and reflected light.

before the time of Euclid (approx. 325–265 BCE); the Parthenon, for instance, was built a hundred years before Euclid. Masterpieces of practical geometry had proliferated for a very long time. Having said that, the fresh-minted information from the systematic study by the Greeks must have seemed awesome with its logical purity and the refinement it brought to construction.

Euclid, a first-generation member of the Alexandrian school, was also a Platonic academician. He was both a mathematician himself and an editor/collator of the work of others. The whole direction of his thirteen-volume magnum opus was toward a rational exposition of the dodecahedron—the form that for Platonists denoted the subtlest "etheric sphere" of creation. For Euclid perhaps his logical geometric journey was accompanied by metaphysical contemplation. If so, luckily for his reputation, he kept quiet about this; otherwise his work may have been far less respected in the educational and academic worlds.

The minimal constructions now used to illustrate Euclid's work highlight how mainstream geometry, like number, has become wholly "quantitative," deductive, and cerebral; whereas I hope that "qualitative" aspects of both geometry and number will emerge conceptually and graphically in what follows. They are inescapably revealed when we come to the parts of this work dealing with music, where, as we shall find, numbers touch the soul.

THE SAND RECKONER AND THE STARCUT DIAGRAM

Sacred Geometry of the Starcut Diagram is about this particular geometer's meandering journey and findings, many of which will be new to most readers; and quite a few, hopefully, new to all. I have included material in a lighter vein, too, illustrating the way that the subject has lent itself to esoteric riddles and enigmas. Recently it has even reincarnated as a newly invented suite of games (see below). It is a subject that turns up in all sorts of arcana—factual, fictional, and playful.

The name for the pattern that is my main theme has also meandered. When I first came across it I called it "the matrix"; but when the film of that name appeared I felt that the name was somehow taken. Then five years ago, for the purposes of a lecture to the Research Into Lost Knowledge Organization (RILKO), I made up the title "the Sand Reckoner's Diagram"; and that became the working title of the first edition of this book. It makes the point of the simplicity of the figure that can be (and, I believe, was) so easily drawn in the sand, as it was for the picture on page xv. The name was stolen from Archimedes, though his sand reckoner was not a geometer but rather a mathematician, proving that numbers could be large enough to define the quantity of sand grains in the entire universe. It happened that when the Roman army occupied Archimedes's home city of Syracuse, one of their soldiers came across the famous man contemplating a construction drawn in his sand tray, and Archimedes called out: "Don't disturb my circles!" This angered the soldier who thereupon ran Archimedes through and killed him—despite orders that he be spared and captured. So the great man was "sand reckoning" right up to his last moments.

From that first public lecture and the many others I have given since, the sand reckoner name, and a little of my associated work, has seeped out and has even traveled as far as Japan on the internet. I notice that my name has not traveled with it. Alas for the acknowledgment of sources!

Nowadays I simply call the device the Starcut, because of its shape, and that is how it appears in the title of this book, which also picks up on the name that I have given to some glass bead games that I invented—almost accidentally—after all but finishing this book. Readers may know of Hermann Hesse's famous work *Magister Ludi* (*The Glass Bead Game*) where the playing of an elegant, complex, and enlightening game is symbolic of a quasi-mystical educational path. As we shall see, there are aspects of the main symbols in this book that resonate with such a theme. My studies over the years have in some ways been like playing glass bead games, and a recent conversation about just that caused me to actually invent a whole suite of them. They are very simple yet remarkably subtle, and I have no doubt that interested players will invent variations and new games of their own. There is huge potential in the connective nodes of the lattice. A copy of the games' rules are in the appendix at the end of this book.

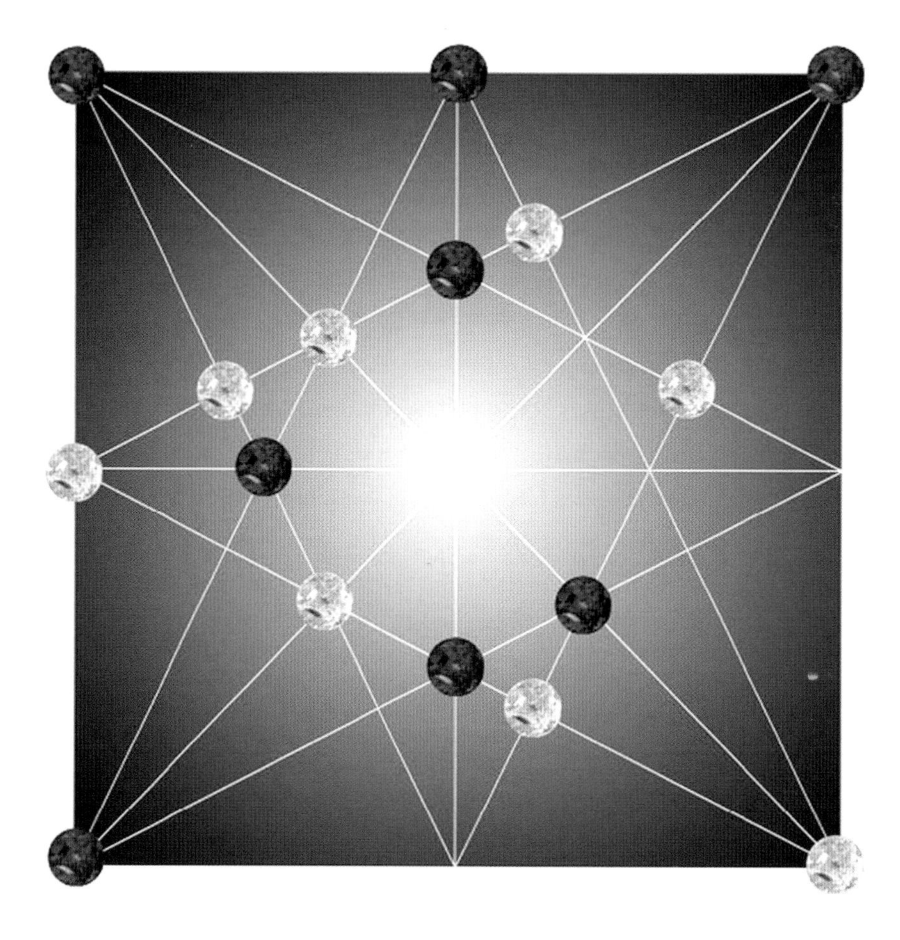

Figure 1.16
The Starcut as a game board.

The diagram itself reveals intriguing relationships between number, geometry, cosmology, and musical harmony. It has not had much public exposure in the literature on the liberal arts, except in the work of the Danish Masonic geometer Tons Brunes. In his vast two-volume study *The Secrets of Ancient Geometry and Its Uses,* the diagram is featured along with a number of other figures. It opens up such a rich terrain that for all the immensity of Brunes's work he does not touch any of the areas with which this book deals.

THE TETRAKTYS

The other pattern, equally "beadlike," that figures in this book, and that has had—strangely, in view of its sacred associations—surprisingly little public attention, is the Tetraktys (fig. 1.17).

Professor Keith Critchlow has published on it, and it lurks in some works on Kabbalah and other esoterica, but it seems that not much else has surfaced since Theon of Smyrna in 100 CE and Iamblichus a couple of centuries after him. Quite a bit of what follows about the Tetraktys has, so far as I know, never been published before.

Having said that, none of us can safely claim originality anywhere in a subject such as this. We simply do not have the historical sources to check out all that has been found before us about a subject that is demonstrably at least five thousand years old, and who knows how much older even than that? It is sobering to recall Alfred North Whitehead's comment: "Everything of importance has been said before by someone who did not discover it."

However, as mentioned earlier, there is a special joy when we discover and understand things for ourselves even if ancient and surer footsteps may have gone the same way before us. And my thesis is that they seem to have done just that. It is unprovable but highly plausible that the Starcut diagram was the graphic source of some of humankind's earliest insights into number and geometry. As will be shown, the sexagesimal number system itself may have come from it—as it certainly could have done.

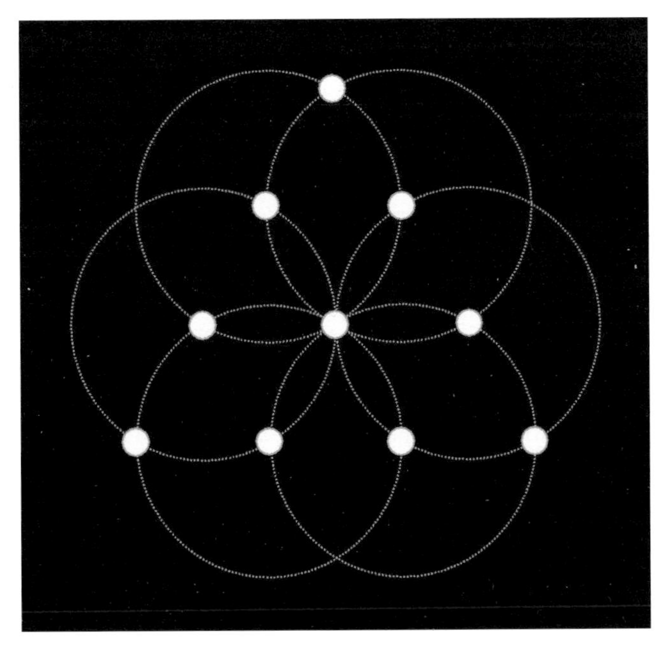

Figure 1.17
The Tetraktys.

STRUCTURE

This book's chapters are fairly self-contained. They are essays and annotations on particular aspects of the main theme, linked together informally as seemed appropriate. Their effect will, I hope, be cumulative. Some information is repeated in different contexts. This helps to highlight the connections and allows the material and the argument to fill out as we go along. Within this subject connections seem rather to form a lattice of points about a volume or perhaps the diverging paths of a maze such as the one at the end of this chapter—with each junction presenting various onward options.

SOME NOTES ON STYLE AND CONTENT

I am indebted to my friend Allan Brown for some fine contributions to the graphics in this book. The apple and blossom in figure 1.2 of this introduction are his. In this subject the visual aspect is *as important* as the verbal. Verbal analysis and visual recognition, we are told, integrate as "bicameral insight"—which is a brain-science way of saying that a picture is worth a lot of words. I have also included illustrations, often preceding the chapters, from design work on which I was professionally engaged while the material herein grew and in which I tried to apply, or simply play with, what I was learning. The diagrams are often far more pictorial than is the custom in books containing geometry. In this way I have sought to illustrate observations and ideas, rather than to present them in the technical manner of textbooks.

The reader will not find any geometric "proofs" but, unless indicated otherwise, the geometry is exact and tested. I made a resolution not to include the sort of typography that one sees in mathematical textbooks. Rarely will the reader find drawings marked with *A, B, C,* and *x, y, z*—and nowhere is there *<ABC* = *<XYZ,* and so on.

I have not recompiled material already much published in the field of sacred geometry, though some repetition of material is inevitable to ensure that readers, for whom the subject is new, get basic information. Compilation of detailed facts and figures in this field can have the effect of somehow closing the whole topic down—leaving the reader with the sense that all is known and there is nowhere left to go. It isn't, and there is. One can read a hundred associations with some number or figure and yet be left with no intuition of the deep connectedness behind the associated material. Who tastes knows.

REASSURANCE

Many people, myself included, with memories of schoolroom traumas, recoil from numbers and any kind of analysis of geometric shapes. So it has surprised me that my work has in fact led to some new mathematical/geometric results. It turns out that returning to basics can still do this. And they certainly are basics. The book is about very elementary forms that really can be drawn and appreciated with simple tools and concepts: an ordinary square, a couple of ordinary triangles, a few stars, some familiar musical harmonies, and a range of easy numbers that fall into recognizable patterns. If things get too advanced, my mind becomes like a sand tray with too much drawn in it—confused and fuzzy at the edges. As someone who continually failed mathematics examinations as a child, I will hardly be a very demanding guide!

MISTAKES

There may be mistakes. I think there are none to do with the key information or the accuracy of the geometry, but, such as they are, they are all my own. The Islamic injunction that any artist or artisan should make a deliberate mistake because only God is perfect, always seems in excess to requirements—imperfections pop up anyway without any contrivance on my part. And there will be gaps too. In a huge subject like this, one can only bear in mind Ezra Pound's implied caution to all writers presuming to deal with "factual" matters:

> *I can remember*
> *A day when the historians left blanks in their writings,*
> *I mean for the things they didn't know . . .*
>
> Ezra Pound, Canto XIII

Figure 2.1

Painted nineteenth-century Tibetan mandala of the Naropa tradition, Rubin Museum of Art, New York.

2
Unity in the Liberal Arts

Philosophical constructions should be relished like ambrosia and nectar . . . delight in them is genuine, does not decay and is divine . . . capable of producing greatness of heart. In themselves they do not make us eternal, but they bring us to a sure knowledge of eternal natures.

IAMBLICHUS, *PROTREPTICUS*

THE MNEMONIC IS THE MESSAGE

A mnemonic is an aid to memory. Whatever else they may be, the Starcut and the Tetraktys are both mnemonics, as are figures such as the Kabbalists' Tree of Life, or horoscopes, or the mandalas of Buddhism, or the Sri Yantra of the Saivites and a host of other sacred emblems. Sometimes the veneration paid to such devices so overwhelms their mnemonic use that what started as a figure intended to enlighten and inform becomes a further layer of mystification.

The Tree of Life shown in figure 2.2 on the next page appears in many forms. *The Jewish Encyclopedia* states that its earliest form had eight spheres and was the invention of the Ikwan-i-Safa, or Brethren of Sincerity (also known as the Brotherhood of Purity)—Shia Muslim scholars in Basra in the tenth century. Their eightfold device was

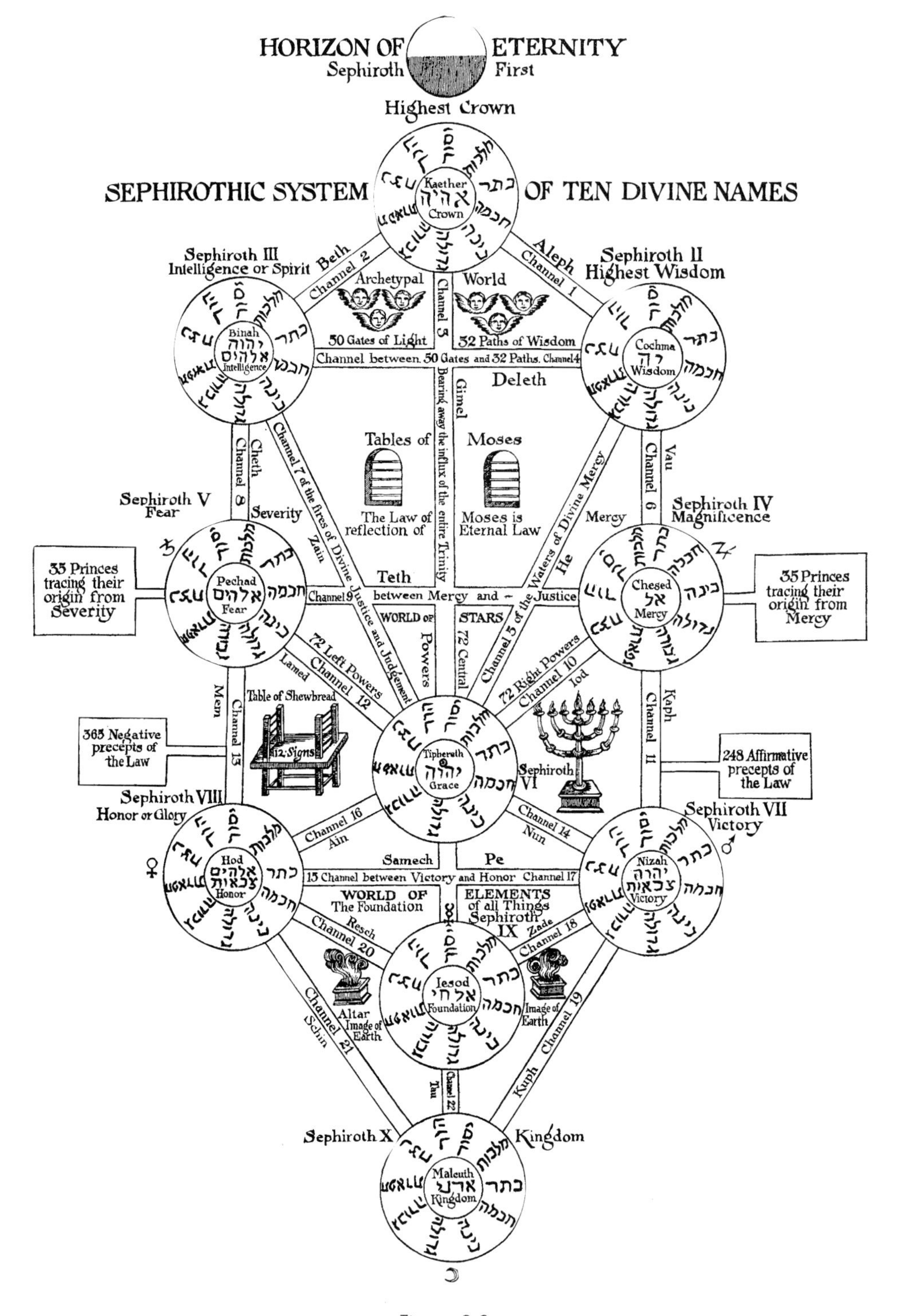

Figure 2.2

The Tree of the Sephiroth (spheres) from the Kabbalah.

developed into the tenfold Tree, which took account of other streams of ancient Hellenic and Judaic thought. The Tree is said to contain the principles of Being and the pattern of Creation from the ineffable into manifestation; it is proposed as the key to the soul's journey of return to its source.

This mnemonic has often been used to carry such huge amounts of information that it has become something of a mystery in itself. The version here includes, in addition to the ten spheres and the paths between them, the numbers of the trumps of the tarot and the glyphs of the planets. So there are at least three entire mnemonic systems packed into this single figure.

The mandalas of Mahayana Buddhism are also maps of a journey—this time conceived as a path "inward" to the transcendent center of awareness. They too are mnemonics rather than objects of worship and

Figure 2.3
The Tibetan Kalachakra mandala.

are considered to encode all the information necessary to guide the aspirant through complex practices of visualization, mantra, and meditation within a comprehensive path to enlightenment.

THE LIBERAL ARTS

Sacred geometry embraces all the seven classical liberal arts derived from the traditions of Plato's Academy. Three of these arts—grammar, dialectic, and rhetoric—later became known collectively as the *trivium* or "the three roads." Those terms can be decoded for general understanding as knowing the language (grammar), knowing how to think logically (dialectic), and knowing how to communicate eloquently to others (rhetoric). Part of the art of rhetoric was certainly the art of how to remember. As was written in *Ad Herennium,* an anonymous first century BCE treatise: "Let us turn to the treasure house of inventions, the custodian of all parts of Rhetoric—Memory."

The great orator Cicero is known to have used a *loci,* or "placing" memory method, since referred to as "the Roman room," as one of five skills needed for effective rhetoric.

Four of the liberal arts, later called the *quadrivium,* "the four roads," refer not to the skills of the student, as do the trivium, but rather to the four fields of study that Plato identified as proper to an educated mind: geometry, number, harmony (including music and architecture), and astronomy. As Plato wrote:

> Every diagram, system of numbers, every scheme of harmony and every law of the movement of the stars, ought to become clear to one who studies rightly.

Studying rightly in these fields means the "artist" will need to employ the skills of language, thought, memory, and expression enshrined in the trivium.

Of course the languages and logics of these different fields are not simply verbal. Music has tonality, rhythm, melody, and harmony. Architecture involves structure, proportion, mechanics, and much else.

Figure 2.4

Here we see represented grammar, dialectic, rhetoric, astronomy, geometry, arithmetic, and music. Plato was rumored to have had the words "Let none ignorant of geometry enter here" written on the portal of his academy. The symbols above, carved on the portico of the Burrow Library in Memphis, Tennessee, implicitly allude to that injunction.

Number has to do with aggregates, ratios, and numerical operations. Geometry is a language of spatial relations, surfaces, angles, and limits. Ancient astronomy was first a language of naming and recording bodies, judging calendrical cycles and periodic relationships, through which events could be dated and predicted. Each of these languages has its logic and its rhetoric. A guiding aspect of my study has been how these languages share common themes on common ground.

To remember the three skills (trivium), the four fields (quadrivium), and their interrelationship, I find it helpful to use, as a mnemonic, the first of the five of what I call "spherics"—the tetrahedron. It is shown in the center of the group pictured in figure 2.5 on the next page. It is the geometric abstraction of the form emerging at the second multiplication of cells in the womb. With deep intuition, Plato, in his *Timaeus,* has it as the second stage in the unfolding of the World Soul.

The spherics are often described as the "Platonic solids" these days—or, more technically, as the regular polyhedra—but the older term *spheric* refers to their essential quality as being regular forms fitting perfectly into the sphere, and inside which another sphere can fit, perfectly touching the center of each face. They don't have to be solids, they can be considered as wire-frames, standing waves, or whatever; they are the regular lattice-potentials intrinsic to the sphere itself; also, their spherical geometry was known and used, in practice if not in theory, anywhere between a few hundred to a couple of thousand years

Figure 2.5
At the top are the dodecahedron and the icosahedron, in the center the tetrahedron, below are the octahedron and the cube.

before Plato, turning up as furrows on many carved round stones that have been discovered in Scotland (see fig. 2.6).

Like the other spherics—the dodecahedron, the icosahedron, the octahedron, and the cube (as in fig. 2.5)—the tetrahedron has other associations that we will touch on later. For now we take it simply as a mnemonic. It's a mnemonic with an extra message in its geometric structure.

Imagine the four faces of the tetrahedron as four doors through which one can enter the volume. Each face may be taken as one of the quadrivium: number, geometry, harmony, and astronomy (fig. 2.7).

Figure 2.6
Nearly four hundred such sphere-stones have been found, almost all in Scotland. Archaeologists dispute their use. The author speculates that they may have first developed as weights for tying into river nets. Many were found in riverbeds, however many highly skilled examples, such as this one, incorporating tetrahedral spherical geometry, show that they became decorative objects in their own right.

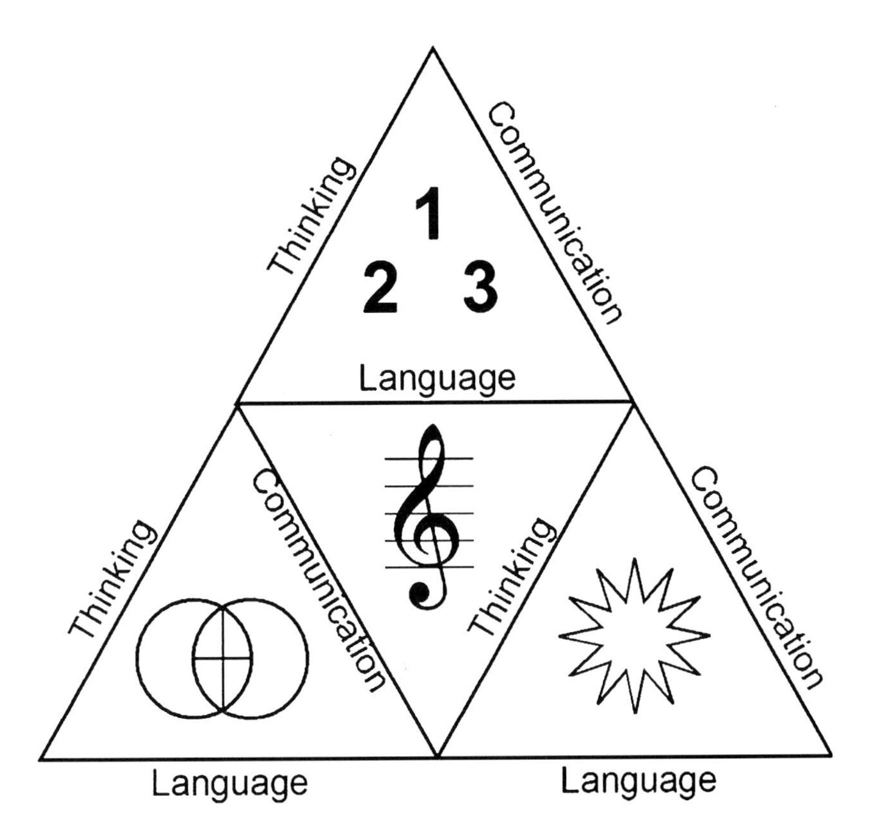

Figure 2.7
The tetrahedron is here opened out into its component triangles illustrating the four arts, each bounded by three skills.

We go through a door with a key piece of information appropriate to it. Then we find that the key that let us through that door also fits the other doors. So we may go in the door of number, and find that we can emerge from a different door—for instance, that of music. Very simple examples of this might be to say that the first development of number from 1 to 2 denotes equally the musical octave; or, in geometry, the relationship of the radius to the diameter of a circle; or the equal light and darkness of an equinox. As we shall see there are many far subtler concordances than those. So, if the various doors all open with similar master keys, what is it that is inside the space itself? Answer: a *cosmos,* which, to the Greeks, denoted a "beautiful order" recognizable as a unity that transcends the separate fields in which its manifestations arise.

Figure 2.8

Allan Brown's graphic reflection on the interconnectedness of the quadrivium.

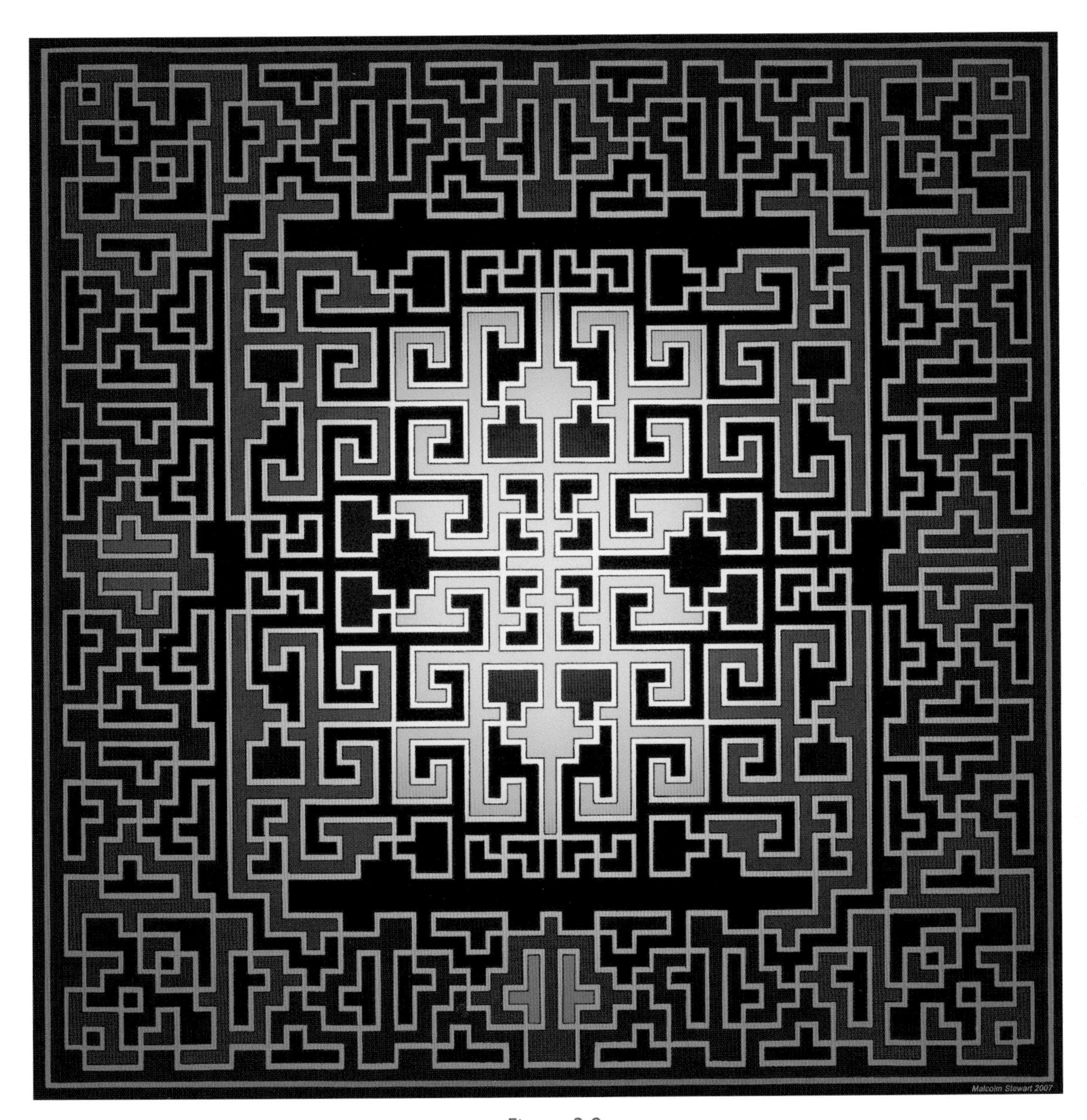

Figure 2.9

In this maze, one may start anywhere on the golden line that meanders at right angles. The game is to find the unique strategy that takes one everywhere along that line—but only once—and returns to the original point of departure.

OF LONDON

3
Opening the Box

Some symbols always mean the same thing.

ALAN MOORE, *PROMETHEA*

In architecture, city planning, furniture, pavements, game boards, wall tiles, plaid designs, graphs, grids, and weaves—the geometric square is everywhere. One speculates that knowledge of the square form goes all the way back to some ancient ancestor who first wove rushes or scored lines in the sand. Certainly by 20,000 BCE or so a Solutrean cave artist had inscribed square figures on the wall of the Cosquer Cave in southern France (see fig. 6.4, p. 64).

The square as a mathematical function is everywhere too, as the multiplication of a number by itself and, in square root form, as the unique factor that, multiplied by itself, produces a given number. Squaring is a feature of formulas for areas of circles and spheres; for geometric theorems; for coordinate systems; in the inverse square law; in laws concerning vacuums, gases, gravitation, radiation, acceleration, and more. Laws of mathematical logic and laws of nature are full of squares. Einstein's famous $E = mc^2$ is only the most famous equation that involves squaring. In that connection it was once commented to me that whereas some people are most interested in E (energy), others in m (solid matter), others in c (speed and light), I seem to have got stuck with the bit that most people

Opposite: Figure 3.1 Trafalgar Square.

hardly notice—the little number “^{2}”. Squaring, however, is the “power” of a number, as defined by mathematicians. The ancient world, too, had it right: “qualitatively” the square denotes the manifest result that comes from the creative dynamism of the triad. Squaring adds magnitude to simple length. Whenever one sees the symbol √ or a superscript number (e.g., $\sqrt{31}$ or 123^{7}) the squaring function is implied. A line of unit length becomes a plane of unit area when squared.

Figure 3.2 is a unit square, that is, the side length is 1. It has some immediate implications. Simply in the drawing of a square, another length, that of the diagonals, is suggested by the relationship of the opposite corners with each other. The diagonal of the unit square is a line that reveals its nature when it is itself squared; it becomes the side of a square twice the area of the original unit square. This new square has an area of 2, as is shown by counting up the component triangles. The line of its side is thus “the root of a square of value 2”—therefore “the square root of 2” ($\sqrt{2}$).

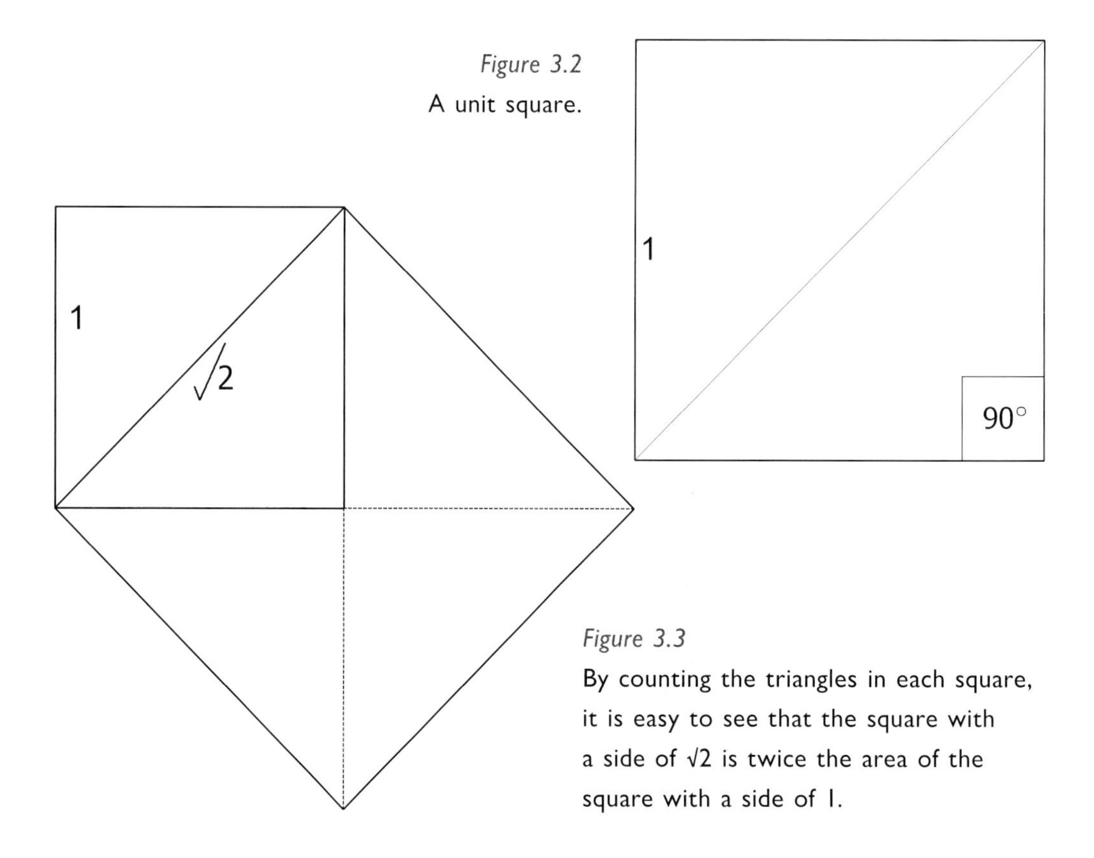

Figure 3.2
A unit square.

Figure 3.3
By counting the triangles in each square, it is easy to see that the square with a side of √2 is twice the area of the square with a side of 1.

There is an immediate numerical mystery. The actual length of this diagonal produces the decimal 1.4142135 . . . , a number that goes on interminably and never falls into a repeating pattern. This means that the decimal part cannot be expressed as an exact fraction because it never settles down. A fraction is a ratio between the number above the line and that below. It is said to have been Pythagoras himself who first realized that √2 was neither a whole number nor a fractional ratio and was therefore "irrational," meaning literally "without ratio." It is reported also that Pythagoras only taught his private group (the *mathematikoi*) about irrationals and the related qualities of the pentagon—which, as we shall see, fairly seethes with another irrational: √5. Apparently one of the private group talked indiscreetly to some of the public group (the *akousmatikoi* or "hearers"). Versions of the story have Hippasus, the indiscreet student, being put to sea in an open boat, banished, or even murdered—which one doubts—by Pythagoras himself.

Though the Pythagoreans may not have known it, that diagonal and its awkward length was known more than a thousand years before in Mesopotamia, from where we even have a clay tablet that gives a solution of √2 correct to five decimal places!

Figure 3.4
Cuneiform tablet circa 1800 BCE. Mathematical scholars are agreed that the writing gives a series of fractional measures that combine to give an accurate result for √2 way beyond that which could be seen by visual measurement. How the ancient world developed the theoretical knowledge implied here remains a mystery.

Returning to basics, what actually is a square? Can it be dismantled beyond its obvious familiar and rather bland face? It certainly can, and this book is the outcome of just that.

Despite my grumbles about the way geometry was taught in school, we had a mathematics teacher, Mr. Ronald Hunt, who taught us something very important. At the top of his voice he would shout the word "WOMEK!" The word was an acronym from the initial letters of "write or mark everything known." It is a key to finding fresh information. When we do this we are much more likely to pick out potential developments latent in the geometry. When, for instance, in order to get a perpendicular bisector, we draw a full vesica, rather than mean little arcs, we are "womeking."

We can apply this to our main topic, which is simply an ordinary square—a flat box with four equal sides. Drawing such a square we may give each side some chosen length—1 being the simplest. Then having marked the right angle and put in the diagonals it might seem that we have finished. But in fact we know far more than this because a square is itself a construction. It has an internal rationale that unfolds in simple steps from the vesica that is the root/womb of all geometry. In its unfolding, it carries information with it. The sand reckoner's square is generated by seven similar circles that establish nine points. One only notices this if one "womeks" the process by fully drawing everything in as it becomes relevant. It can be drawn very simply, by rotating a forked stick in the sand and stretching and snapping down a cord for the straight lines; such methods have been with us from very ancient times.

Reading the elements of the construction in figure 3.5 in sequence we get:

1. A vesica—two equal circles, two centers, two intersections;
2. A resulting right angle makes a new center;
3. A centered third circle cuts the right-angle lines at four points;
4. These points become centers for four new circles, making four corner points;
5. The points connect, giving a star lattice in overall square form.

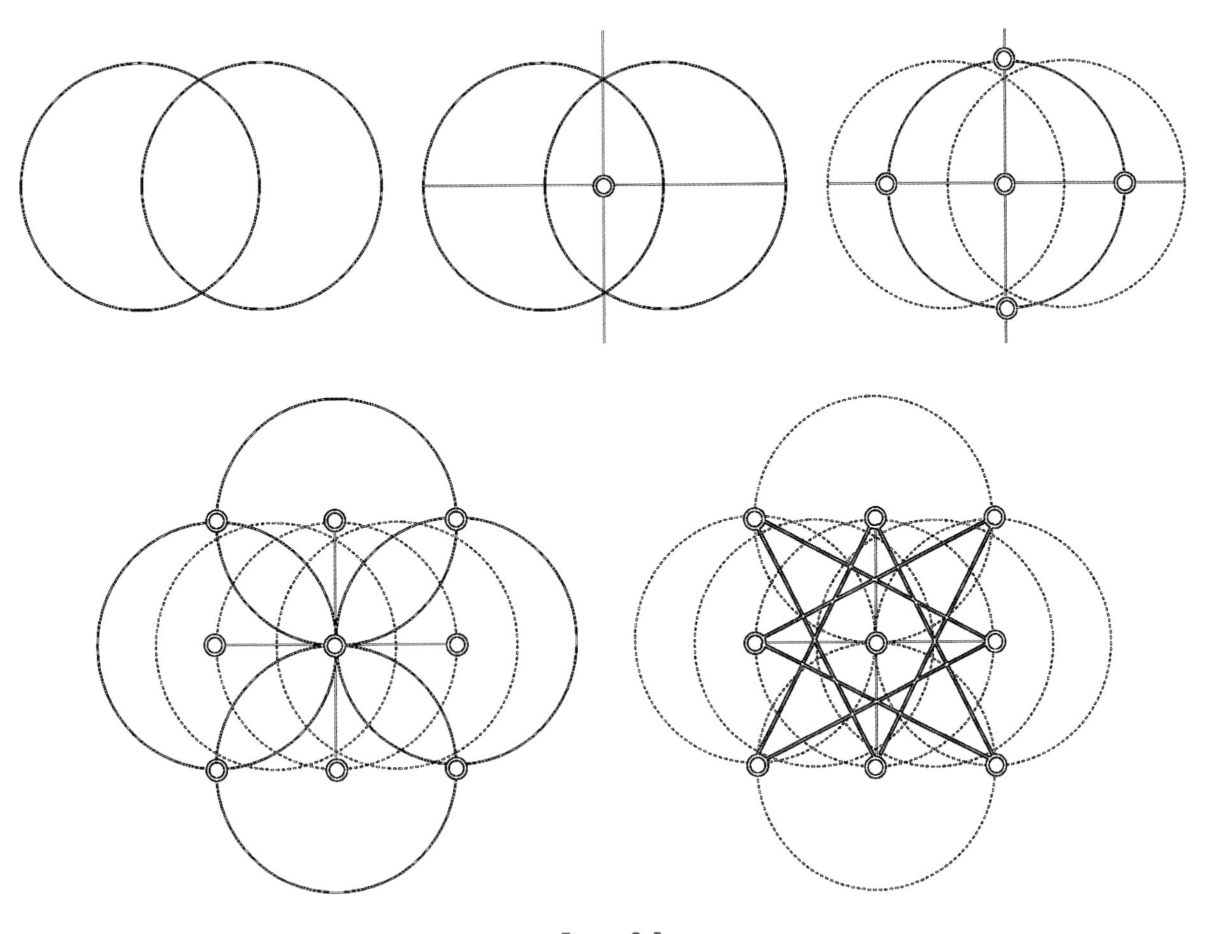

Figure 3.5

The square (and the Starcut) constructed from seven similar circles.

It is now as though we have lifted the surface of our square, like a hinged lid, to see its ancestral inner workings. In actually making it one has learned far more than is immediately obvious from just looking at it as an equal-sided box shape.

The lattice triangulates the square in a way that is the key to all that follows. The triangle is the basic straight-line plane. Two points establish a line. Three points—unless all on the same line—establish a triangular plane. A square, of course, can be cut into triangles in an infinite number of ways; but here using the guide points from the square's own generation, we get a natural, economic, and uniquely elegant triangulation that is truly loaded with information. There are two forms of the

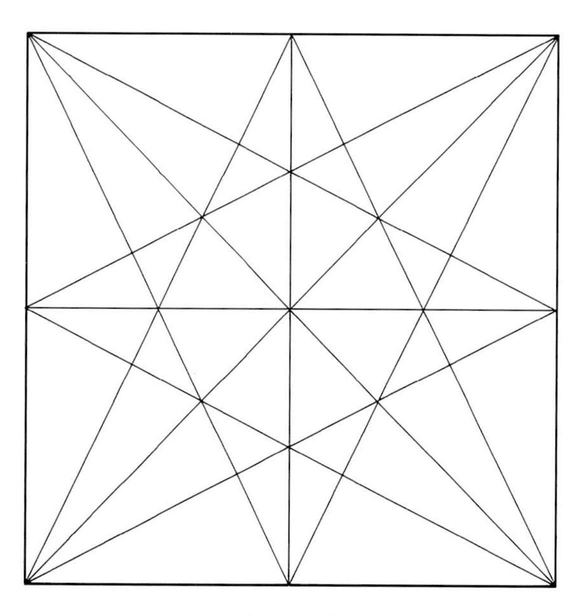

Figure 3.6
The Starcut lattice
within the square.

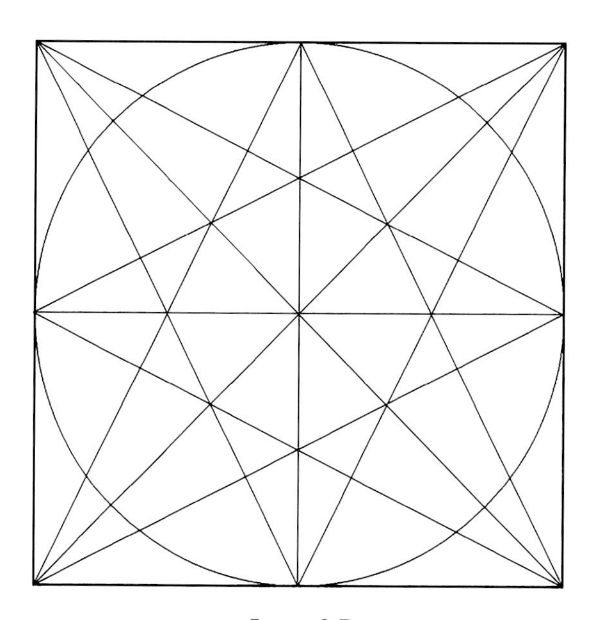

Figure 3.7
The Starcut in the central circle
that contains it and that is
part of its construction.

lattice (fig. 3.6 and 3.7); sometimes it is useful also to retain the central circle that was part of the construction (fig. 3.7).

It works as a mnemonic device for a vast swath of lore from the metric and sacred geometry of India, Mesopotamia, Egypt, China, Greece, Judea, Islamic culture, and Renaissance Europe. What is more, if one forgets its content, one has only to observe its measures, numbers, and proportions and everything reveals itself again.

I first came across it by chance as a possible grid underlying an Indian temple ground plan, of which more in the next chapter.

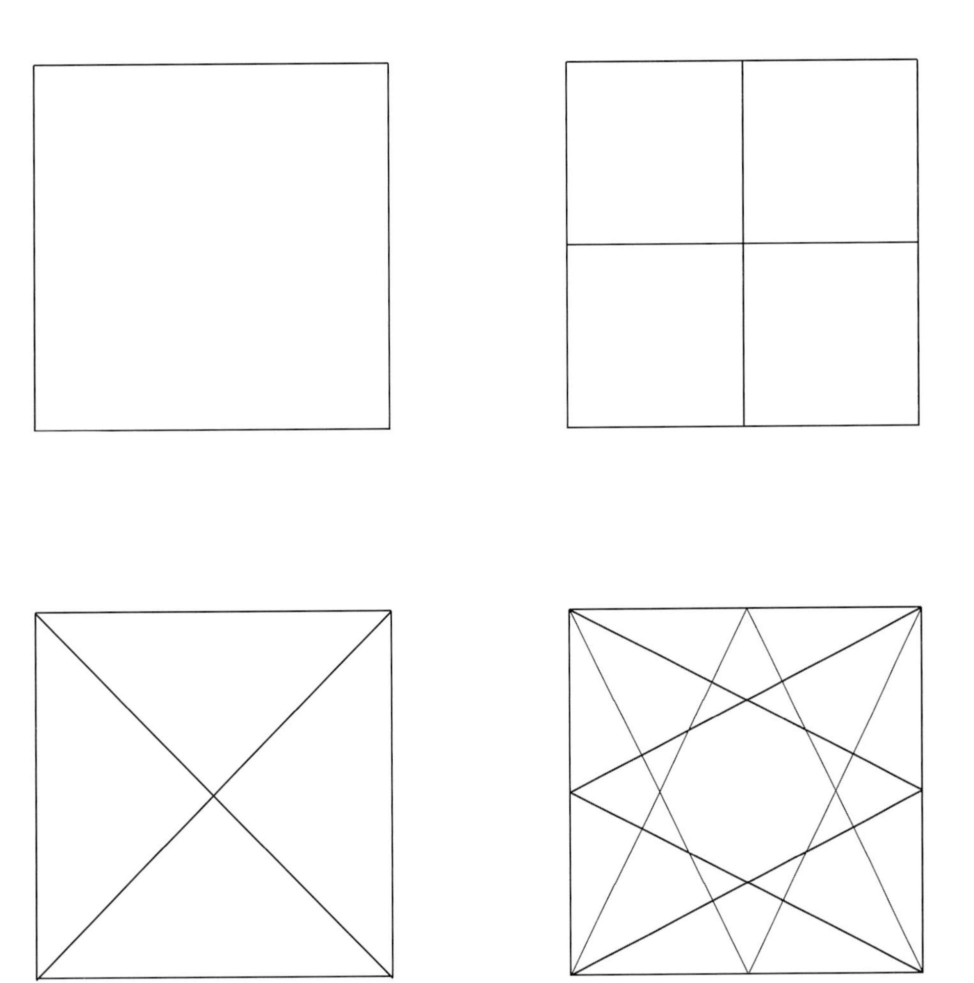

Figure 3.8
The elements of the Starcut are the square, the meridians, the diagonals, and the mid-side diagonals.

Figure 4.1
The Shore Temple in Mahabalipuram, Tamil Nadu, India.

4
Altar of the First Fire

Simply to think of building temples is to be freed of the sins of a hundred incarnations.

AGNI PURANA

MYTHIC ASSOCIATIONS— THE SWAMI AND THE ARCHITECT

A form of the Sufi tradition that particularly cherishes the practice of the "Holy Names" was first introduced to me as if in disguise. It was a path that seemed to weave through the more arcane aspects of all traditions, rather than staying solely within Islam, which is undoubtedly its home faith. I met a man in Sri Lanka who, while not a geometer himself, served as an inspiration in that direction for me. I at first took him to be some sort of mage, until friends told me he was a Buddhist *anagarika* (homeless mendicant), then others said that he was a Siva swami, and that was what he later came to be called (Sivakalki Swami). He taught me to respect the heart of all traditions. This approach is characteristic of what another more famous sage, Coomeraswami, called "the unanimous tradition"—a somehow more emphatic and gutsy term for what is known more airily as the perennial philosophy. After meeting this Siva swami, who surprised me by suggesting I read the Koran, I, as an ex-Catholic

Figure 4.2
Az-Zahir, "the Manifest."

priest, felt freer about exploring and drawing on teachings from many sources. In the study of "sacred" geometry—a way of thought that really is a common ground in almost all the great spiritual cultures—this has been a most helpful approach. One thing that this enigmatic man-of-all-teachings taught me was that one of the "Ninety-Nine Most Beautiful Names of God," derived from descriptive terms used in the Koran, is the name az-Zahir: "the Manifest" (figs. 4.2 and 4.3).

There is a tradition in Jewish Kabbalah that features the same word. Here the Zahir may be an object, a person, an animal, an event, or an abstraction. It may indeed be anything—because its nature does not lie in the form it takes but in the peculiar effect that it has upon the person that perceives it. The effect of the Zahir is that it is unforgettable—once encountered it can never again be put from the mind. Perhaps this association goes way back into the ancient meaning given in the *Hebrew-Chaldean Lexicon*: "Zahir—brightly shining, that about which one should be warned." It has also been said that at any time in history some particularly potent example of the Zahir exists in some form or other lying in wait to entrap the attention of the unwary. Jorge Luis Borges, the Argentinian author who wove a tale around this folkloric theme, claimed that at one time the Zahir was a seemingly insignificant coin and at another it was a particular tiger.

All this may only be a haunting fiction but it felt as though I had encountered just such a phenomenon in the form of this simple square diagram; and it turned out to be highly appropriate since, as already mentioned and as we shall see further, the fourfold square has, from earliest times and in all cosmologies, been a symbol for that which is materially manifest (the Zahir) in creation.

The late Sivakalki Swami gave me a name—following a custom common in the East. It was the name of the Vedic deity Agni. I did

Figure 4.3

Plaque of the "Ninety-Nine Most Beautiful Names of God" from the Holy Koran. Calligraphy by the late Siddiq el Nijoumi, master potter; concept and design by the author.

not use it socially but thereafter I did research any reference to Agni that I met. Agni, in the Vedas, represents the "etheric fire" of creation, the first fire of the dawn, the flame on the sacrificial altar, and the first spark of spiritual motivation in the human heart. Agni can be creative or destructive. There was plenty to reflect upon!

Two years later, back in the UK as a BBC TV producer on the hunt for program ideas, I got to know Professor Keith Critchlow, the architect, who was at that time teaching at the Architectural Association and at meetings of the Research into Lost Knowledge Organization (RILKO) in London. Through him I was inspired to extend a longtime interest in patterns and mazes into a detailed study of the vast lore of geometry and proportion that is connected with sacred and monumental sites, symbols, and buildings. It was from Professor Critchlow that I first truly appreciated how the study of this form of geometry opens one of the greatest and most universal of all the metaphor systems that people have ever used to embody their ideas and intuitions about the unity and harmony of the cosmos.

This double influence of the swami and the architect resulted in one of my early geometric investigations being into the temple plan that I mentioned at the end of the previous chapter. It was the ground-plan mandala of the sacred fire altar—the altar of Agni (fig. 4.4). I received a copy of it on a visit to Professor Critchlow.

The ground plan's first—most obvious—characteristic was that it was a square with three differently sized sets of internal tiling (fig. 4.5). At that time I had no knowledge of the background culture of such a design; but simply looking at it and counting its tiles I noticed something about its numbers. They reminded me of numbers that were already familiar because I had read John Michell's seminal, enlightening books *View over Atlantis* and *City of Revelation* where he distills a "canon" of numbers that, in simple or multiple form, are to be found in ancient monuments, such as Stonehenge, and in visionary descriptions, such as that of the New Jerusalem in the book of Revelation, or Plato's description of the measures of Atlantis in the *Critias*. I had also begun to explore the numbers of the Platonic lambda and the Tetraktys. These will feature later in this book.

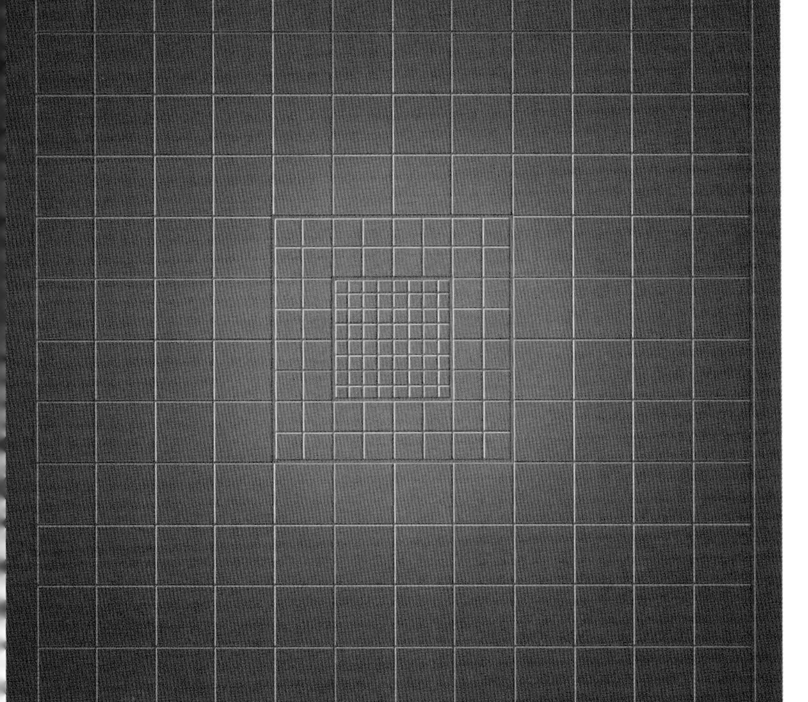

Figure 4.4

The ground plan of the sacred fire altar of Agni.

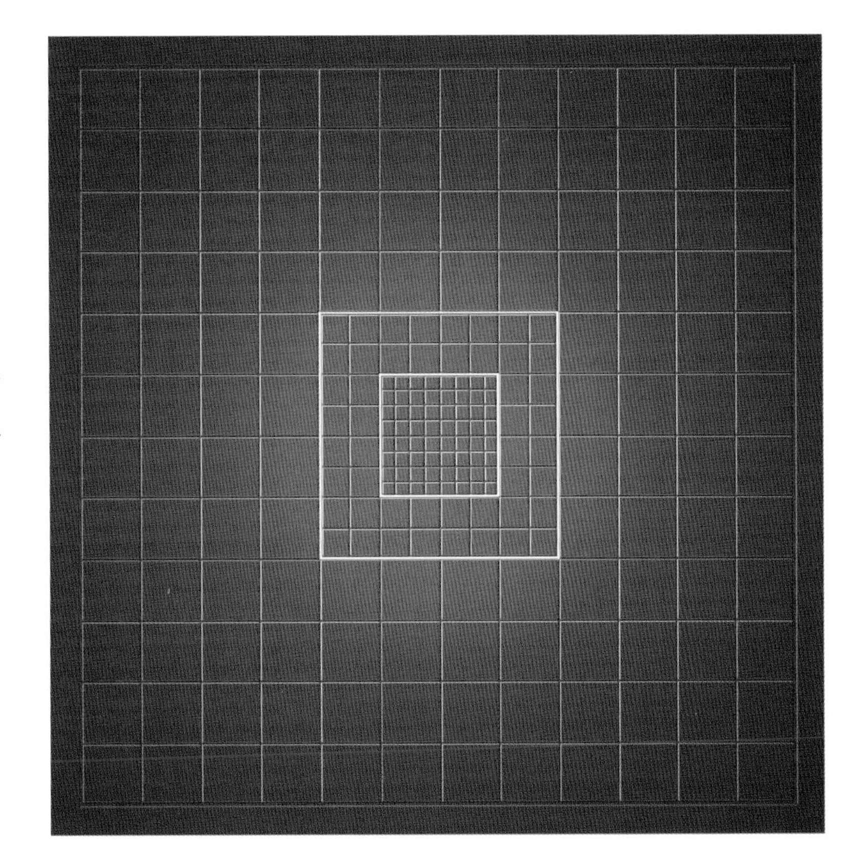

Figure 4.5

The tiling system highlighted.

I counted up the tiles and noted their numbers under arbitrary headings: the precinct (the big square), sanctuary (the medium-sized square) and altar table (the smallest square). I had no way of knowing if these were the correct distinctions. At that time I did not know anything at all about the ancient Vastu architectural system from which this diagram came. The system is reliably said to be toward five thousand years old, dating back to Vedic times. As does the Agni name and concept itself.

The entire precinct suggested an underlying checkerboard measuring $12 \times 12 = 144$. From the Bible and from John Michell's studies, this was a familiar number. But the central 16 tiles of the 144—what I thought of as the sanctuary area that was $\frac{1}{9}$ of the overall square—had been differently tiled. This part had been filled in with quarter-sized tiles. So there were actually only 128 checkers of the largest size. The number 128 was also a familiar number to me because of work on the Tetraktys.

The central $\frac{1}{9}$ had been marked out $8 \times 8 = 64$. This too had turned up in Tetraktys studies, and of course inevitably recalled the 64 hexagrams of change in the *I Ching* and the squares of the chessboard. Of these 64 checkers, 48 were apparent (48 also was a notable Tetraktys number) while a further 64 small squares covered the very central area, the altar table, which is actually $\frac{1}{36}$ of the entire ground plan.

I played around with some of these components, observing that if the numbers of the most densely tiled area were applied to the whole square, the sum total would be 2,304—also a number that I recognized from John Michell's work; as too was the total of the smallest tiles that would occupy the mid-sized $\frac{1}{9}$ area—that total being 256—a further notable Tetraktys number, and the number sometimes called "philosophic C," because this was a traditional frequency to which a piano's middle C was tuned until the early twentieth century when pitch standards were generally agreed upon to suit orchestral wind instruments.

If any readers find that all these numbers and associations induce a sort of delirium I should say that, at this stage I was simply groping about without any unifying rationale. A bundle of numbers with haphazard connections into a handful of different cultures in different historical periods was presumably just some numerical fluke with no particular significance. Later on we will return to these and other numbers. And to those who fear numbers—please don't worry—we will be looking for patterns of coincidence rather than doing calculations.

At this stage in my first investigation I decided to see if the ground plan might reveal more when decoded by means of the tools of traditional geometry: the straightedge and compasses. Though unskilled in mathematics I had started, by this time, freelancing as a stained glass designer. Geometry was good.

THE STARCUT

So with square and compasses I tried to discover the most likely manner in which the ancients may have formed the ground plan on the land, using only poles and cords. Inevitably I followed the vesica construction that we have already seen. First the site's orientation would have been established according to the dawn sun; Agni's fiery associations, of course, connect directly with this.

Keith Critchlow had published the method used for this as recorded in the *Manasara Shilpa Shastra*. Using a cord, a circle is drawn around a standing pole. The shadows of the rising and setting sun of (for instance) the winter solstice fall upon this circle, and the points where the shadows touch the circle are marked. A line through those points indicates east and west. These points, or other appropriate points on the east-west line, are taken as the radius for a vesica piscis, from the right angle of which north and south are derived (fig. 4.6, next page).

Having literally "oriented" the first guidelines for the ground plan, the makers would lay out their grid on real terrain. In theory, the thing could be laid out by simple measurement, marking distances along the edges; but in practice, I knew from my own large-scale stained glass and mosaic work, that in any fair-sized precinct the right angles can easily drift. Normally a square form is confirmed by checking that the diagonals are of equal length and that they intersect at their midpoints and at right angles to each other. Given the likely size of the whole area, however, further useful guidelines might have been derived by connecting the corners of the square with the centers of their opposite sides, to tighten it up.

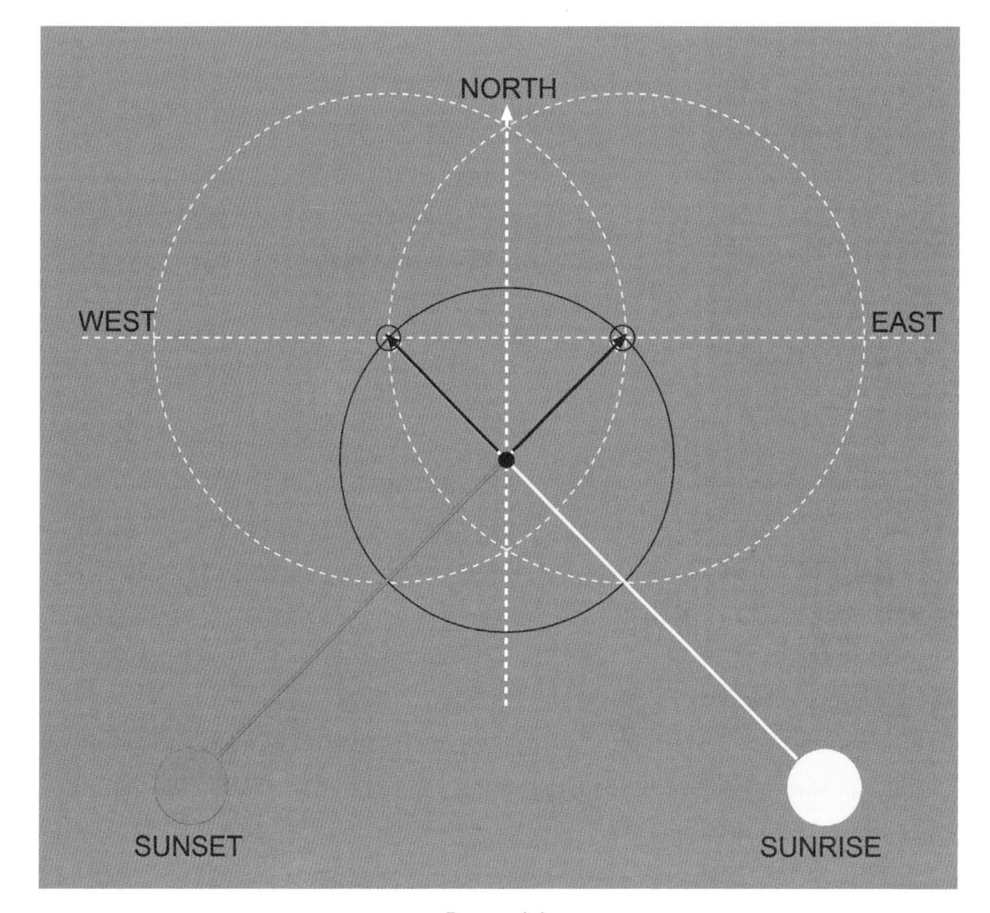

Figure 4.6
The method for determining the precise east-west axis for orienting a temple that is laid out in the *Manasara Shilpa Shastra* using sunrise and sunset on the winter solstice.

This is precisely the construction that we have already seen in generating a square from seven similar circles. It is a construction with the added advantage that while reinforcing the exactness of the structure, it also provides—from within itself—all the measures necessary for the finished design with its internal tiling pattern.

This was when I first came across what I've called the Starcut, or sometimes the "sand reckoner's diagram," because it is so easily drawn on the ground or a sand tray and is, as we shall see, the source of so much numerical and geometric information. It has some obvious geometric properties that suggest that it could indeed have been used in the con-

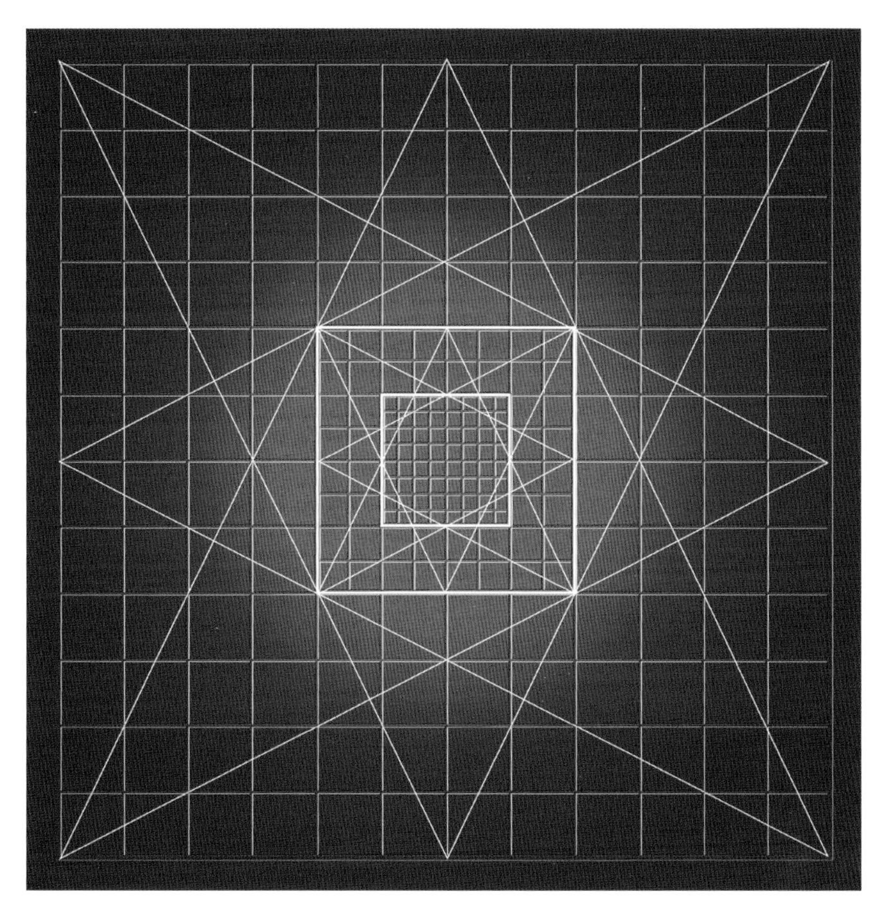

Figure 4.7
The three precincts of the temple overlaid with two Starcut diagrams.

struction of the ground plan for the Agni altar. Though I knew that I could be wrong as to its historical relevance—the exercise had yielded a form that had something about it: for me, an unforgettable Zahir.

I did not know then, and still am not certain, whether this lattice was a knowingly applied diagram in Vedic times or whether it is a metaform that happens to be a useful mnemonic for all that follows. We shall find it is an even more effective mnemonic than the tetrahedron mentioned earlier, not only because it packs in far more information but also because the Starcut's information, even if forgotten, can be retrieved, with full detail, directly from its own geometry.

The Starcut did have some obvious properties that meant it could

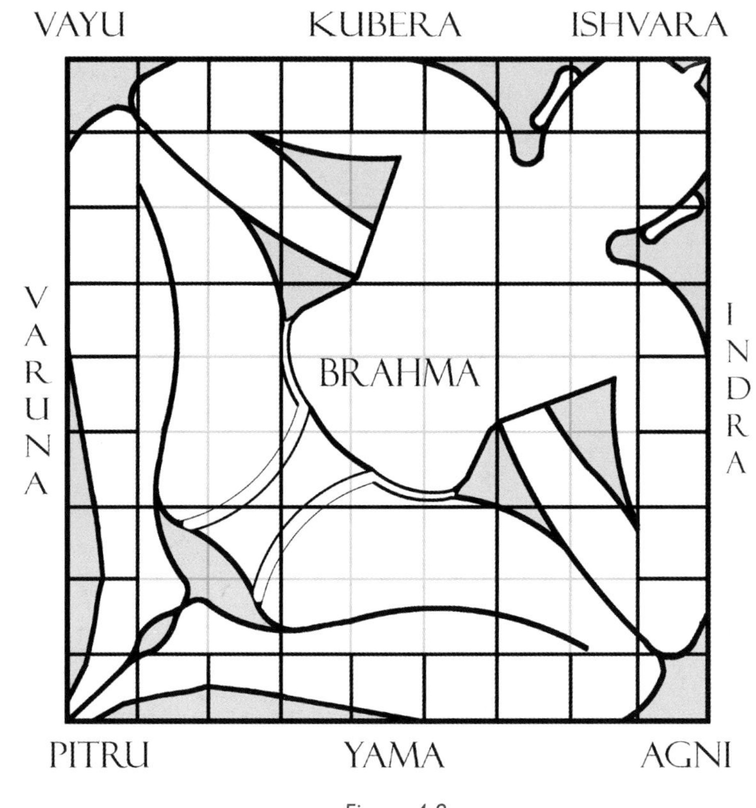

Figure 4.8
The Vastu Purusha Mandala.

have been a practical matrix for establishing the ground plan. For a start, it provided guide points for cutting a square into regular checkerboards of smaller squares, which the Agni plan involves. A second Starcut drawn at the center indicated a ⅑ section of the side of the main square. The division into ⅑ connects with the Vastu Purusha Mandala (fig. 4.8), which, I was to learn, is the fundamental structure of the Vastu tradition of which the Agni altar is one example. Vastu is to the Indian tradition what feng shui is to the Chinese. The Vastu Purusha Mandala is a field of 81 squares (9 × 9). Exactly the same 9 × 9 checkerboard also turns up as the basic square used in Vedic mathematics.

The legend associated with the Vastu Purusha Mandala is that a great formless being threatened to consume the world but was rendered harmless by being pinned face downward on the earth by forty five gods. (Is it just coincidence that he is always pictured aslant at a 45-degree angle?) He was placated by then being accepted as the guardian of all consecrated land, and thus all temples are laid out on his frame. Nine major gods of the Hindu pantheon are ascribed directional positions as seen in figure 4.8—the other thirty-six are also mapped on a complete Vastu temple plan. Agni, being the deity of the first fire of dawn, is in the southeast.

Before looking at this Starcut in detail, it is worth noting that the square itself is the initial object of contemplation for the architect in Vastu tradition, which, although of immense age, is still in use by Hindu architects today. Though I had not realized it—knowing at that time nothing about Vastu—my numerological contemplation of the Agni ground plan was in some ways very much in line with what the Vastu architect does when first establishing the appropriate mandala for a temple. Nowadays in the West, people assume that mandalas are basically circles, but Vastu lore puts this assumption into context. The point, the polygon (in most cases, a square), and the circle are all involved as is evident in the quote on page 51 from Ganapati Sthapati, master modern Indian architect:

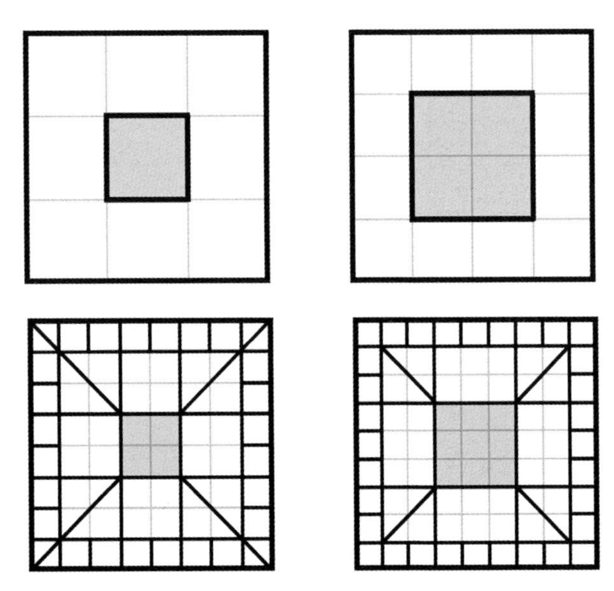

Figure 4.9

Four Indian mandala grids in traditional square form. Reading top to bottom, left to right they are: the Pitha, the Mahapitha, the Parama, and the Manduka.

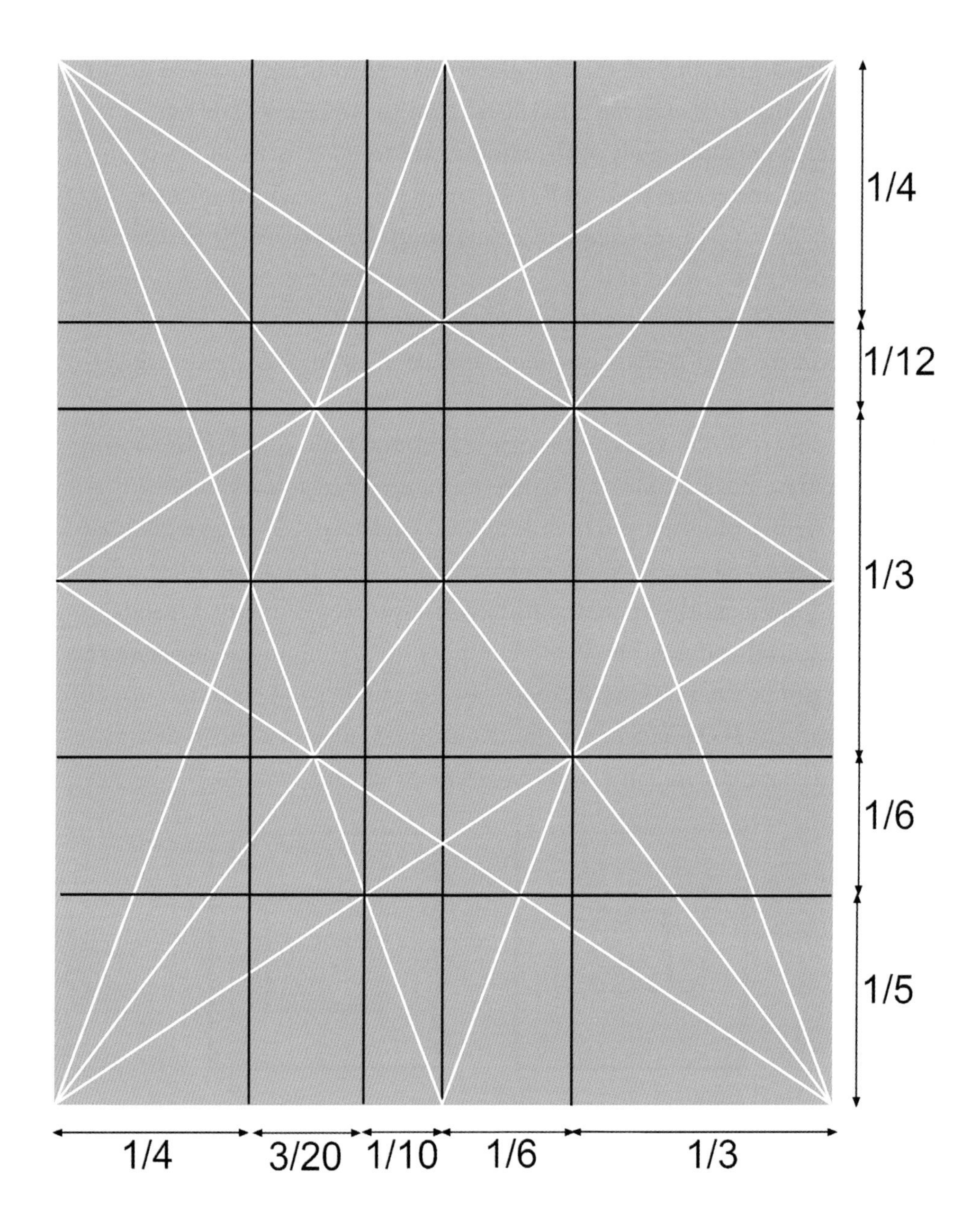

Figure 4.10

A Starcut oblong. The diagram can be drawn in any rectangle. As a stained glass designer I have used it to "make friends" with whatever rectangle a client's window happens to be. The lattice cuts the sides into many measures, of which some samples are shown here. It suggests harmonious foci for whatever design is to be developed in the space.

The architect or Sthapati begins by drafting a square. The square is literally the fundamental form of sacred architecture in India. It is considered the essential and perfect form. It presupposes the circle and results from it. Expanding energy shapes the circle from the center; it is established in the shape of the square. The circle and curve belong to life in its growth and movement. The square is the mark of order, the finality to the expanding life, life's form, and the perfection beyond life and death. From the square all requisite forms can be derived: the triangle, hexagon, octagon, circle, etc. The architect calls this square the vastu-purusha-mandala—vastu the manifest, purusha the Cosmic Being, and mandala, in this case, the polygon. When completed the vastu-purusha-mandala will represent the manifest form of the Cosmic Being; upon which the temple is built and in whom the temple rests. The temple is situated in Him, comes from Him, and is a manifestation of Him. The vastu-purusha-mandala is a mystical diagram. It is both the body of the Cosmic Being and a bodily device by which those who have the requisite knowledge attain the best results in temple building.

Figure 5.1

The Sri Cakra—commonly called the Sri Yantra—incorporates layers of information relating to centers of awareness in the body, the sound vibrations considered proper to them, and an entire tantric cosmology. This is the author's draft for a glass version in a friend's shrine room.

5
The Tartans of Pythagoras

The Starcut, at the nodes where the lattice intersects, gives guide points for dividing the main square into a variety of regular sections. With it one sees how to checker a square by 2, 3, 4, 5 (figs. 5.2–5.5), and by combinations and multiples of those divisions (fig. 5.6 on the next page).

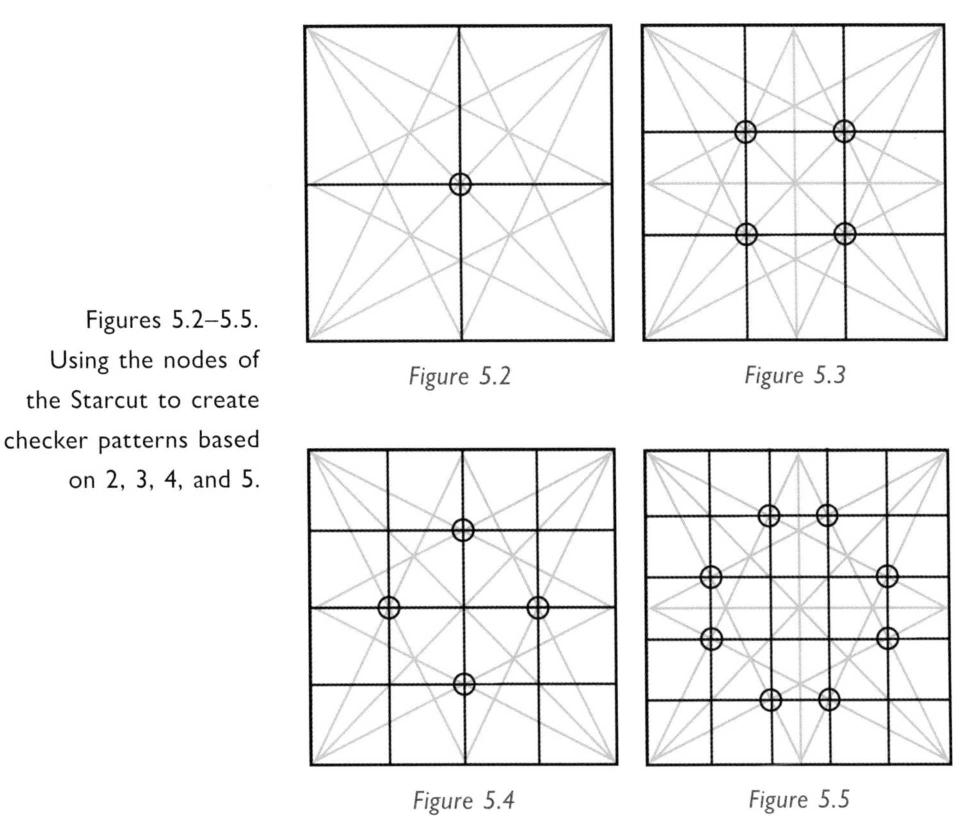

Figures 5.2–5.5. Using the nodes of the Starcut to create checker patterns based on 2, 3, 4, and 5.

Figure 5.2

Figure 5.3

Figure 5.4

Figure 5.5

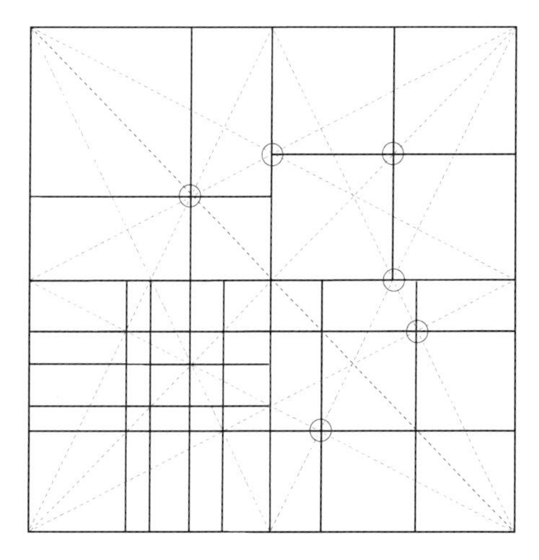

Figure 5.6
The square cut into quadrants (clockwise from upper left) of 3 × 3, 4 × 4, and 5 × 5, with the lower left quadrant combining all three to suggest a tartan checker pattern.

"If Pythagoras had had the good fortune to be a Scot," as the monk said, "he might have found an endless source of plaid designs in this lattice." The square in figure 5.6 is cut into quadrants. Taking them clockwise, 3 × 3, 4 × 4, and 5 × 5 sections are indicated in each quadrant, with the salient points ringed. The last quadrant has these segmentations combined together into a rudimentary tartan checker pattern—further developed and illustrated in figure 5.7.

By progressively developing from those internal nodes, any square can be dissolved into an infinity of measured frequencies. The reader will notice that as lines are drawn through nodes, these lines in their turn indicate new nodes on the diagonals and mid-side diagonals, and these allow for further multiplication.

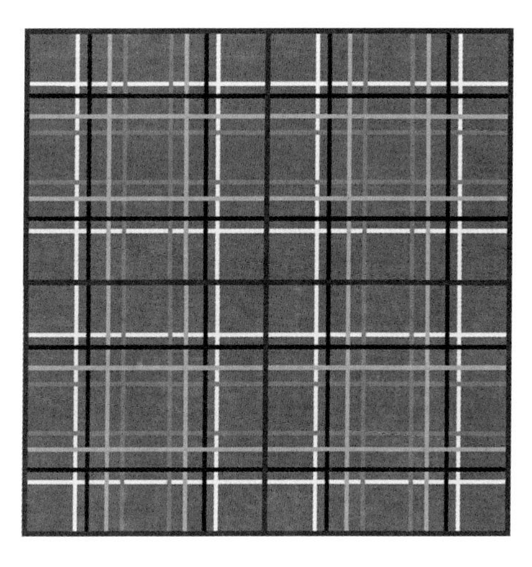

Figure 5.7
The 3 × 3, 4 × 4, and 5 × 5 quadrants combined into a rudimentary tartan checker pattern.

The first test of further applications of the Starcut was in constructing yantras. I had for some years been studying in the Arica School of philosophy whose founder, Oscar Ichazo, has developed many series of yantras as meditation tools. There were over thirty of them current in the Arica materials at that time. They were designed to exact specifications both as to form and color. Some of us reproduced them as paintings, hangings, carpets, and collages for use in different environments and in different sizes. This involved constructing them geometrically, as is the practice in the Indian Sakta Tantra tradition where yantras, such as the famous Sri Cakra, are important mnemonic and contemplative devices (see fig. 5.1, p. 52).

I was interested to see if my sand reckoner's diagram would provide general guidelines for painting the Arica yantras accurately on whatever scale. So I analyzed them using the lattice. All of them were facilitated by it and often in interesting ways. Here is an analysis of just one of them—a fairly simple example (fig. 5.8). It is the yantra used in meditations on the principle of consciousness that Ichazo calls "No Time."

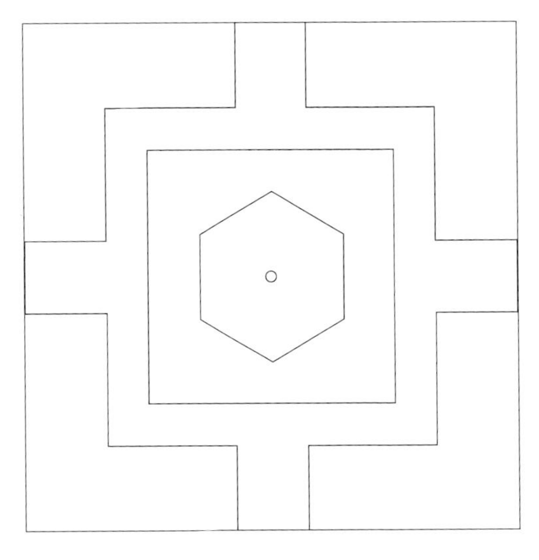

Figure 5.8
Basic outline of the "No Time" yantra.

First, using the ringed nodes as marked in figures 5.9–5.11, the square is divided 12 × 12. The whole square does not have to be so divided, only the areas that are needed for the yantra's outlines. Notice how the lines intersect with the diagonals and mid-side diagonals, developing further nodes as needed.

The lattice is developed to provide a grid that contains all the information that will be needed for the final construction.

The inner box square has sides of a depth that is 1⁄12 of the main square's side length. The circle, in which a hexagon is to be drawn, has a diameter that is 1⁄3 of the main square's side. The hexagon is drawn by stepping the circle's radius six times around the circumference (fig. 5.13).

The compass point is placed at the corners of the box square and the radius is set so as to touch four of the corners of the hexagon, as shown. Arcs are drawn to cut the sides of the main square.

These points on the main square's sides govern the cross form that connects up with the box square.

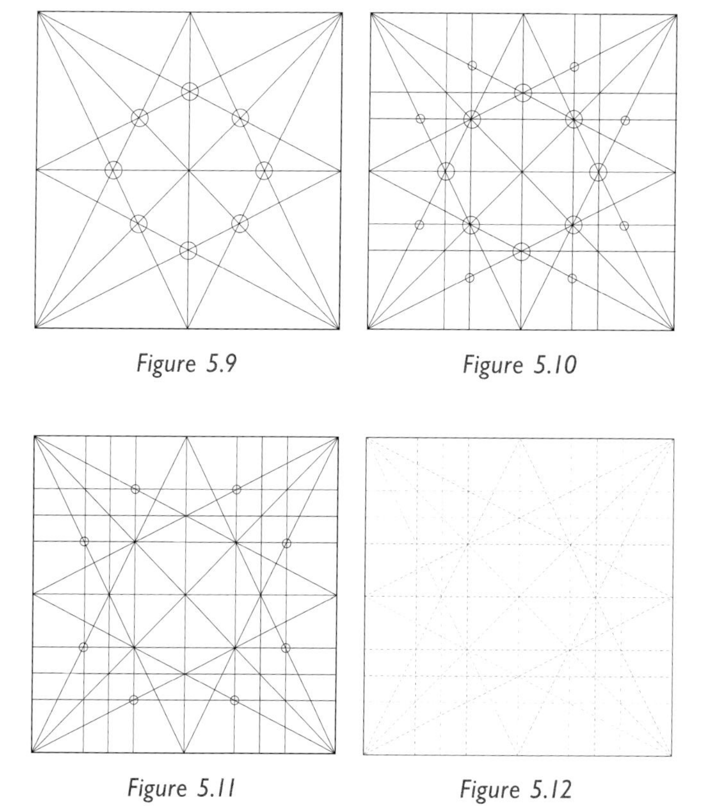

Figure 5.9 *Figure 5.10*

Figure 5.11 *Figure 5.12*

Figures 5.9–5.12. Dividing the square with the Starcut to form the yantra.

The main outline of the yantra is now complete. It only remains to place the central point, which has a diameter that is 1/36 of the main square side. A small Starcut inside the 1/12 checker, divided into three, gives this measure (fig. 5.16).

The yantra is now complete.

I was delighted to find a general solution for constructing the yantras and was amazed at their consistency especially since, so far as I could discover, these yantras had not been designed by the means that I had used in decoding them. This was, however, an obscure finding in an obscure field, but it set me searching.

Other properties of the Starcut pattern started to emerge.

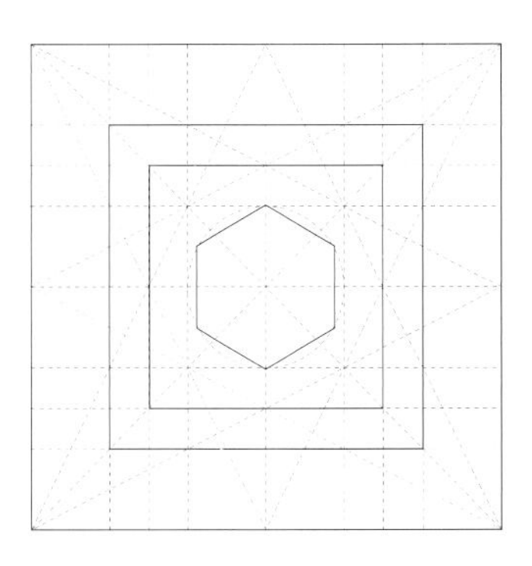

Figure 5.13
The inner hexagon emerges from an inscribed circle, and the surrounding squares are defined.

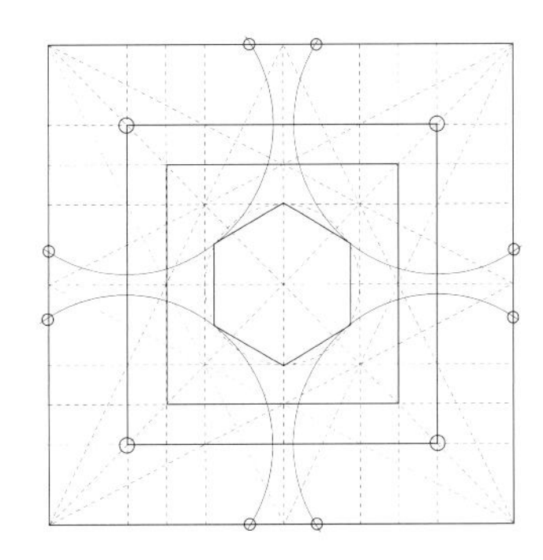

Figure 5.14
Arcs are drawn to locate the arms of the cross on the outer square.

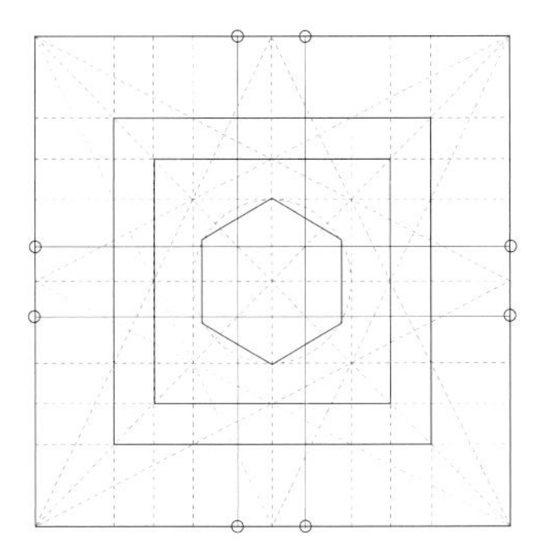

Figure 5.15
The arms of the cross are laid.

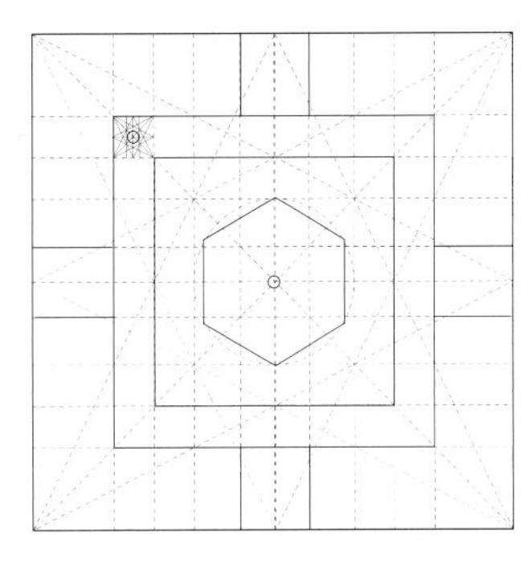

Figure 5.16
The final shape of the yantra emerges.

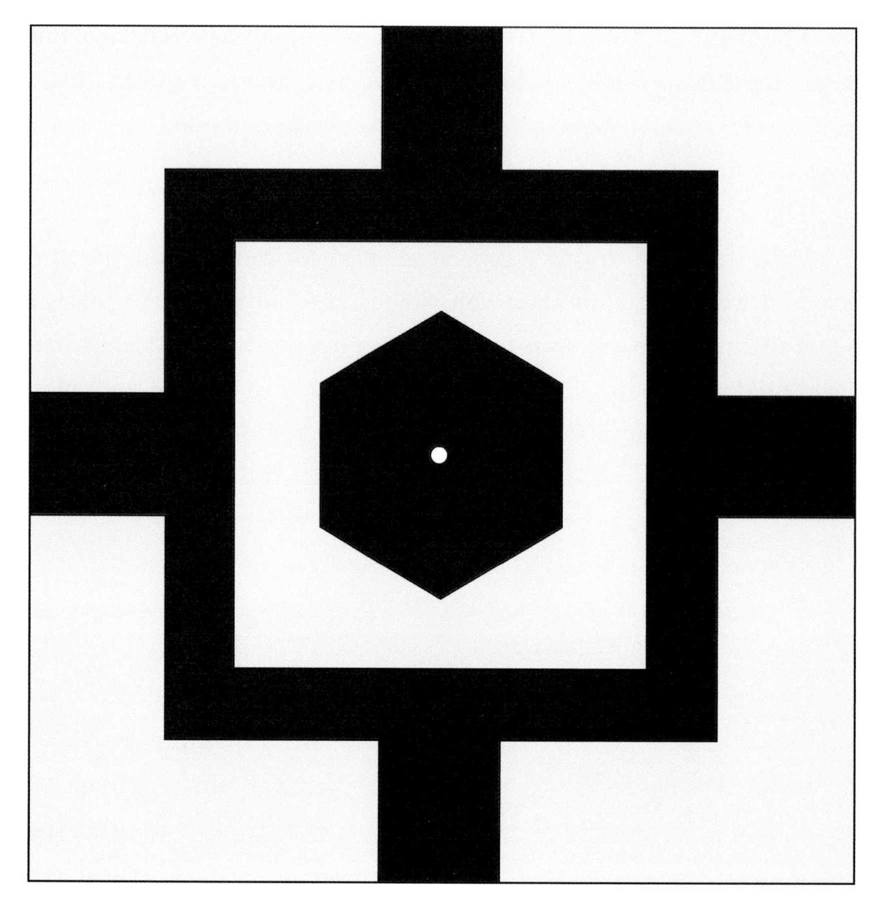

Figure 5.17

Arica "No Time" yantra. The meditation theme with this yantra states: My consciousness is timeless, I am pure present. © Arica Institute, Inc., 1972, 1997; used with permission.

THE WILD BUNCH

The sectioning of the sides of the main square by the method above using horizontal and vertical lines through the nodes turned out to be just the beginning. I thought that after finding how the Starcut would checker a square into 3 × 3, 4 × 4, and 5 × 5—plus their multiples and combinations—I might have exhausted this aspect of its potential. I was quite wrong—further infinities were lurking there!

It was on finding that by using slanted lines through the nodes one seemed to get cuts of 7 and 11 (figs. 5.18, 5.19), that I guessed there might be more in store. But I was unsure because I was working physically with square and compasses and many an inaccuracy can be hidden in the thickness of a drawn line, and emerge to disappoint a geometer who gets too excited by an apparent discovery. And apart from that reservation, there was the fact that division of the sides of the square into measures of 3, 4, and 5, and their multiples, always involved drawing horizontal and vertical lines—not slanted ones—through the nodes of the Starcut, and through further junctions that arose from the new lines one had just drawn.

Compared to that orderly procedure, these sloping lines were mavericks. They cut the sides at angles, and at first I fully expected that they would not stand up to strict measure. Also the numbers themselves—7 and 11—were not numbers that we normally associate with constructions based on squares. I guessed that on trigonometrical examination they would turn out to be inexact approximations. It turned out that I was wrong; they were in fact totally accurate.

It was some time later, in comparing lore from ancient metrology—the study of measures—with this maverick property of the Starcut diagram that there appeared some intriguing connections. Those lines that divide the square's sides into 1/11 and 3/11 sections provided a geometric rationale that happened to marry directly with a beautiful piece of celestial synchronicity pointed out by the late John Michell in his *City of Revelation*. We will see that in the next chapter.

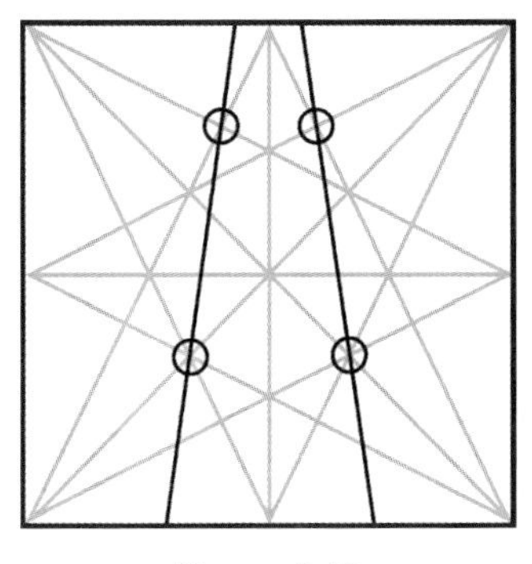

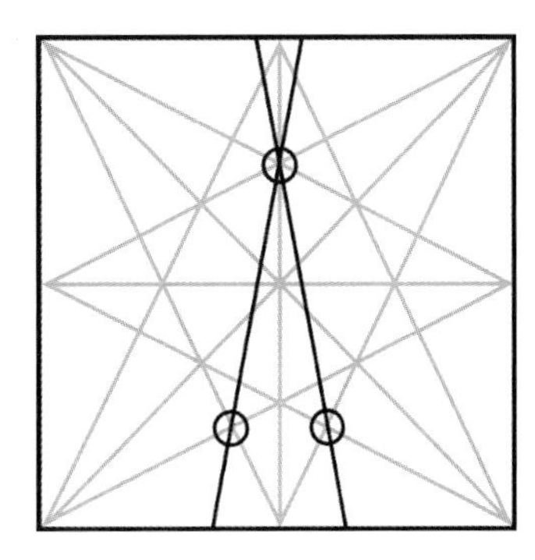

Figure 5.18 *Figure 5.19*

Figures 5.18–5.19. Examples of using slanted lines through the nodes to get cuts of 7 and 11.

6
Pyramid Connections

It is a curious fact of nature, ignored by modern cosmologists but evidently of the greatest interest to their predecessors, that the answer to the problem of squaring the circle is presented nightly to public view, for it occurs in the relative dimensions of the earth and the moon.

John Michell, *City of Revelation*

On page 62, figure 6.2 shows a slightly modified version of the late John Michell's inspired diagram showing the harmony of measures between the sizes of the earth and the moon. In this case, for reasons that will emerge later, I included the profile of the Great Pyramid standing on the base of the square. I have also included the Starcut in the picture, with special emphasis upon three nodes that govern the perfect $1/11$ cuts of the sides of the main square. The sloping lines that pass through the ringed nodes mark off a $3/11$ section of the top side of the main square, and they indicate (if extended through the bottom ringed node) a $1/11$ section in the middle of the bottom line. The black circle represents the circumference of the earth; touching it is the pale circle of the moon. The earth and the moon are both square-boxed, and it is shown that the corners of their respective boxes make a 3-4-5 triangle. John Michell points out that, according to the traditional $22/7$ ratio for

Opposite: Figure 6.1
The Pyramids at Giza.

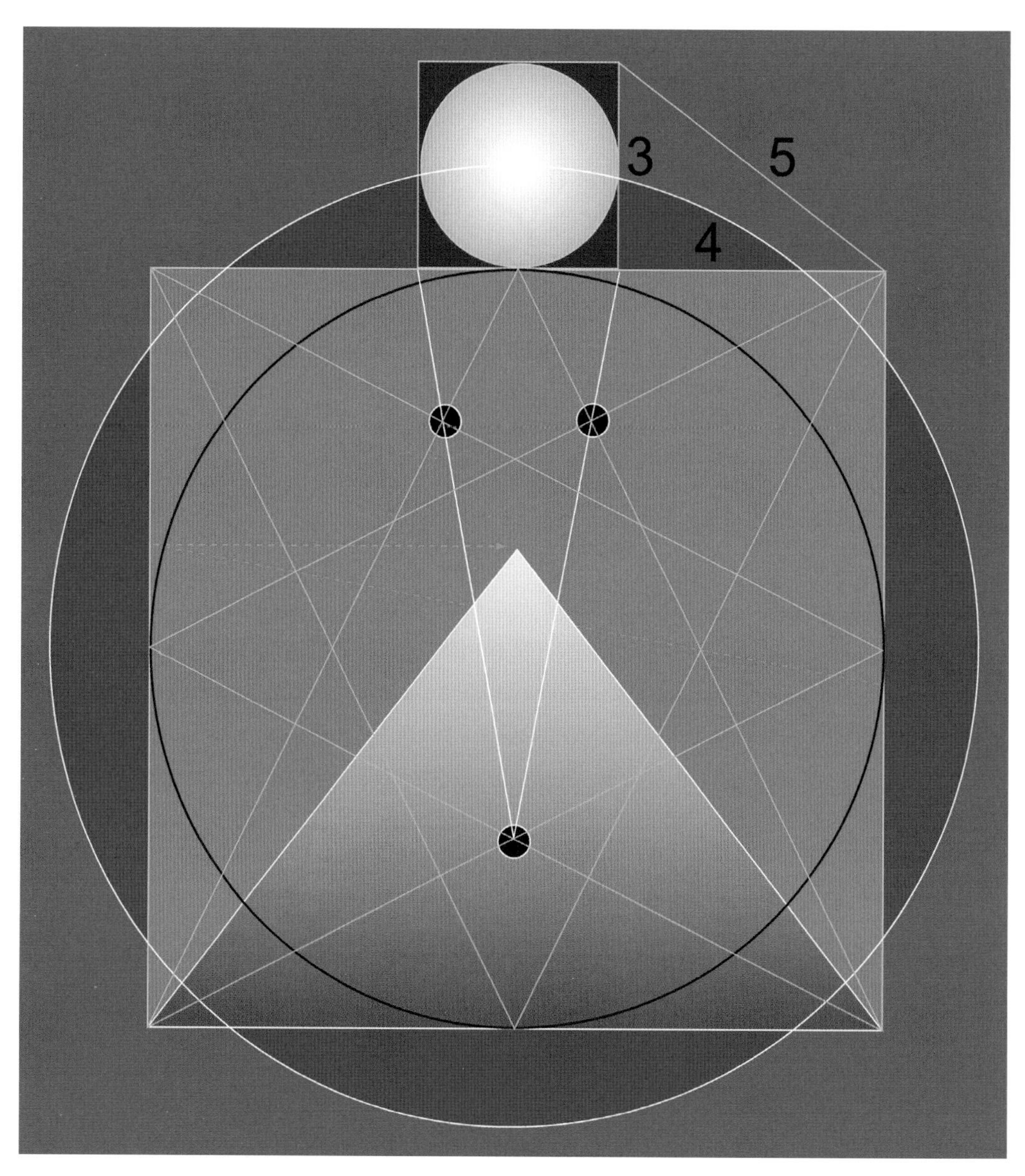

Figure 6.2

Squaring the circle of the earth and the harmony of measures between the sizes of the earth and the moon.

pi (the Greek letter π), if the center of the earth's circle is taken as an origin, and the distance to the center of the moon is taken as a radius, the resulting circle—outlined in white here—is squared by the "box" containing the earth.

What the Starcut adds is the subtending geometry, latent within the square itself, containing exact indications for that $3/11$ section of the earth's square that is also the diameter of the moon. The geometry in this way further corroborates the sense of macrocosmic harmony.

As I studied the Starcut further, the way that it feeds directly into the metrical systems of the ancient world came to be one of its most fascinating and mysterious features. Over the years I have increasingly wondered whether this sand reckoner's diagram actually was a known device of very great antiquity, the contemplation of which suggested primary understandings of geometry and number whose provenance have remained unknown to us. It is such a simple piece of geometry and so closely connected to very elementary constructions.

When we think of the history of geometry we inevitably think of the work of Euclid, but his work came tens of thousands of years after humans had first imagined and employed abstract protogeometric design. This is proved by a piece of engraved ochre (fig. 6.3) found in Blombos Cave on the coast of South Africa and reliably dated as being 77,000 years old. Yes, 77,000! A mere 20,000 years ago the Solutreans—an enterprising people who seem to have crossed the Atlantic from Europe by keeping to the southernmost fringes of the transatlantic icecap of that time—left behind, in the Cosquer Cave near Marseilles, the earliest "square" engravings that I know of (fig. 6.4). Imaginative abstraction is clearly part and parcel of human awareness itself.

It is profoundly interesting that when imaginative abstraction becomes exact measurable geometry, we find that the macrocosmic world actually approximates to our mind's constructions with a remarkable intimacy. Why do earth and moon so nearly equate in a coincidental way, as we saw earlier, and how come an ordinary square anticipates and contains their rationale? The geometer John Martineau's Keplerian investigation of the geometric harmonies of the solar system is well worth looking at (see *A Little Book of Coincidence*). We will see an example of this a little later.

Figure 6.3
Engraved ochre from Blombos Cave, South Africa.

Figure 6.4
"Square" engravings from Cosquer Cave, near Marseilles, France.

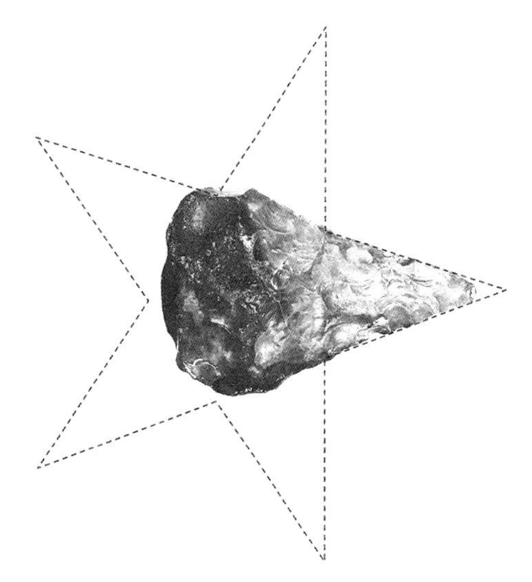

Figure 6.5
Hand axe, Gray's Inn London, flaked and crafted 250,000 years ago. Stone Age spear points, which this resembles, are similarly protogeometric in shape. The angle where the hand would have gripped this is, by a nice fluke, that of a five-pointed star.

Pyramid Connections

The fact that the Starcut gave natural $\frac{1}{11}$ cuts along the edge of the square, taken with the finding that it also gives a $\frac{1}{7}$ cut, immediately set off a train of thought about π (pi), the approximate ratio of the circumference to the diameter of a circle (though $\frac{22}{7}$ is only one of a number of ancient suggestions). This was especially so because, as can be seen from figure 6.6, the $\frac{1}{11}$ section is itself divided in half by the vertical

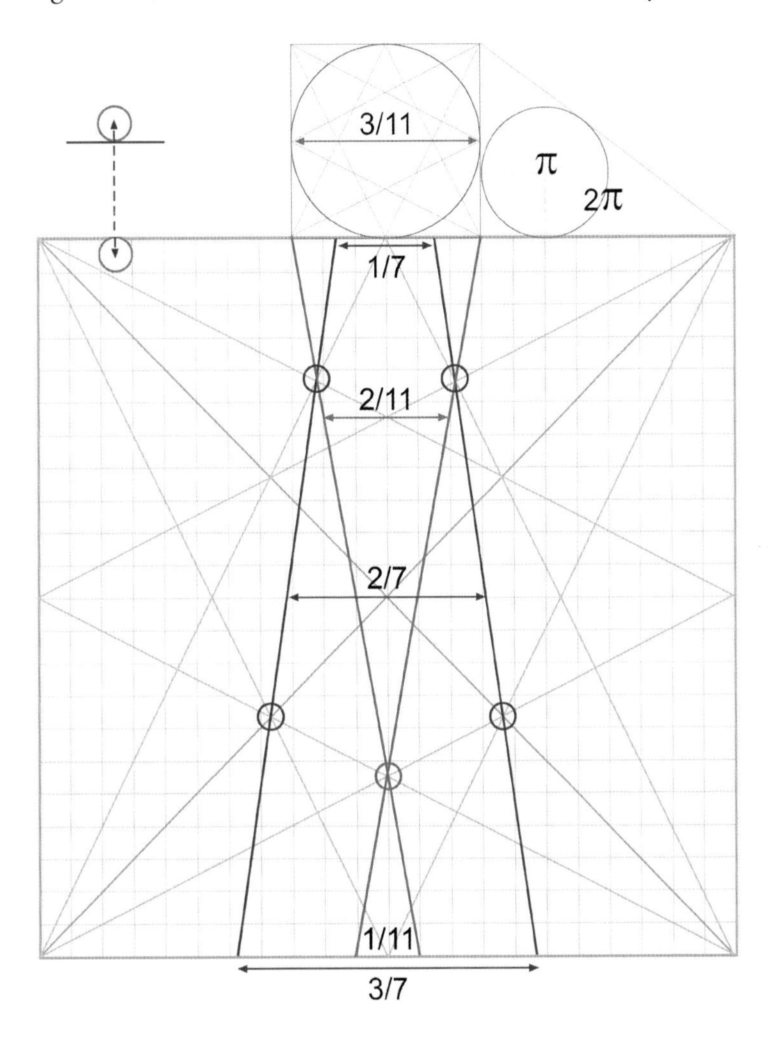

Figure 6.6

The 1/11 section is itself divided in half by the vertical meridian of the main square. Also note the area and diameter of the circle inscribed in the 3-4-5 triangle at the top.

meridian of the main square. So we have $\frac{1}{22}$ as well. The Starcut gives plenty to think about in this connection, and very elegantly. Figure 6.6 does not prove anything, but it draws threads together and stimulates geometric concepts. It was when contemplating this drawing, with the $\frac{1}{11}$ cut of the main square's side length becoming the basic unit of measure, that I first noticed that any 3-4-5 triangle has an in-circle with an area of π and a circumference of 2π.

Pi also occurs in almost perfect graphic form in figure 6.6; one can see at the upper left-hand corner how the circle that fits within the little square that is $\frac{1}{22}$ of the width of the main square, sits upon the line that measures $\frac{1}{7}$ of the main square, which is—according to the $\frac{22}{7}$ approximation for π—the length of that small circle's circumference when rolled out flat. This was the Egyptian symbol for the proportions of the circle; also, some claim, it denoted "eternity." It has been suggested that Egyptian measures accorded with π so well because they measured upward with a cubit stick-measure that was used also as the diameter of a wheel with which they rolled out their horizontal lengths. With such a procedure, π gets automatically built into any geometric proportions used.

The way that the geometries of 7 and 11 and the square and the circle interact gives a sense for how earlier geometers may have explored their relationship. This entire figure, including the segmentation into 22, can be generated from the Starcut's nodes, though not all the construction lines have been included here.

THE GREAT PYRAMID *SHEKED*

The $\frac{1}{11}$ cut led directly to the proportions that are fundamental to the Great Pyramid. It was the writer/surveyor David Furlong who first told me that the *sheked,* the height-to-base ratio of the Great Pyramid, was 7 to 11 (fig. 6.7). This ratio can be read directly from the Starcut, as we have seen, and is further illustrated in figure 6.8.

In figure 6.8 I have also arrowed one of the governing angles of the Starcut that turns out to be the face-apex angle of the Great Pyramid. This is the angle that one would be approaching if one were literally scaling the center of the pyramid's face. It is determined by the

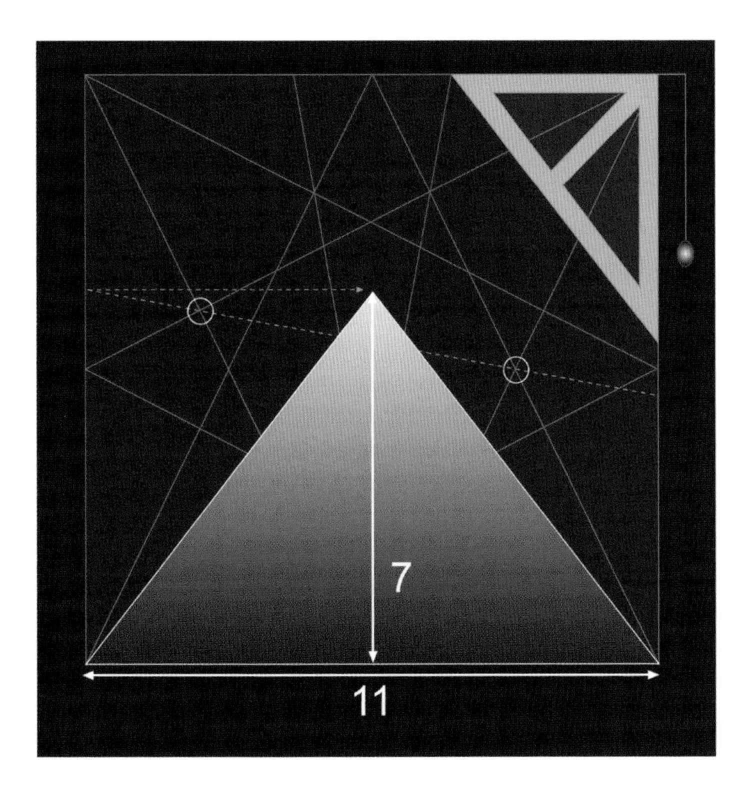

Figure 6.7
The *sheked* (height-to-base ratio) of the Great Pyramid.

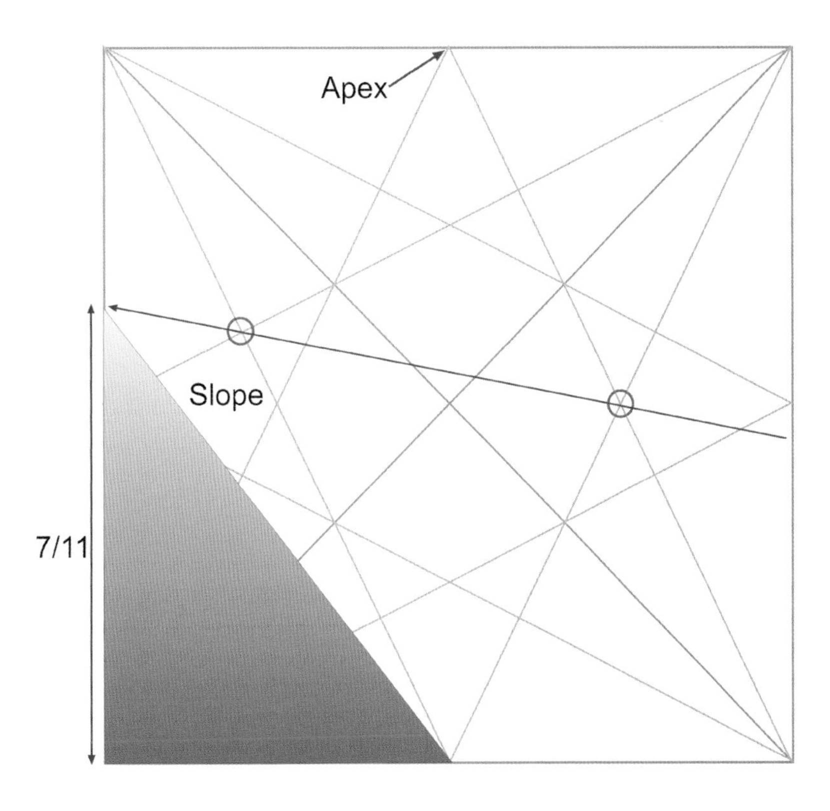

Figure 6.8
The ratio of 7:11 further illustrated.

7:11 *sheked* angle that governs the incline of the converging faces. It is correct to 1/50 of a degree.

Wishing to check how close the 7:11 ratio is to impartially surveyed estimates of the pyramid's height (J. H. Cole 1925), I found a difference of four inches. Since the thirteen-acre ground square of the pyramid, according to the same meticulous survey, includes errors up to a little under eight inches, a possible drift of four inches at the pinnacle seems acceptable. Indeed when one considers that there were, in all, over two hundred courses of stone cut, transported, laid, and fitted, with major internal constructions as the mountainous edifice was built, a discrepancy of the width of the palm of one's hand becomes truly astonishing and convinces me that this interpretation of the *sheked* is correct.

So what about the suggestion that the *sheked* was generated from a knowledge of ϕ (Greek letter phi), which in fairly recent times has come to be called the "golden ratio"? In arithmetical terms this ratio can be expressed as $1:(\sqrt{5} + 1)/2$ (or 1:1.6180339 . . ., the latter number being irrational).

Geometrically, when a line is divided in this ratio, the shorter segment is in the same relationship to the longer segment as that longer segment is to the whole line. Those who believe the Egyptians to have used this ratio say that if the pyramid's height is taken as $\sqrt{\phi}$ and the half-base width as 1, then the slope is of length ϕ (fig. 6.9).

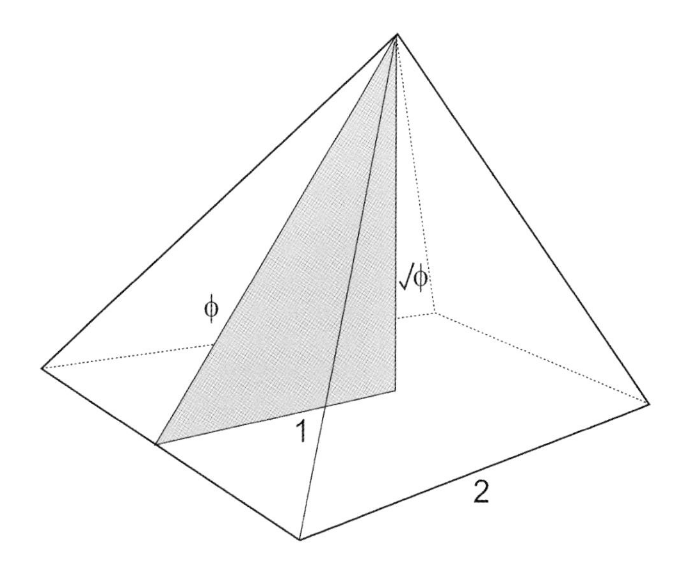

Figure 6.9
If the pyramid's height is taken as $\sqrt{\phi}$ and the half-base width as 1, then the slope is of length ϕ.

Proponents of this view have called the triangle—shaded in figure 6.9 above, with side lengths ϕ, 1, and √ϕ—"the Egyptian triangle," and deduce that the Egyptians knew the golden proportion and were the source of the Pythagoreans' knowledge of it. We can compare these two *sheked* versions by dividing the square root of ϕ by 7 and multiplying the result by 11:

$\sqrt{\phi}$	=	1.272019 . . .
$\sqrt{\phi}/7$	=	0.181717 . . .
0.181717×11	=	1.998888 . . . , so let's call it 2!

In terms of proportions therefore, the two solutions are effectively the same. I suspect the Egyptians derived their measures from the simpler numerical approach. We know they were very good at squares—the base of the Great Pyramid proves that. It was fantastically accurate with no more than an eight-inch error in a thirteen-acre area. We also know that they worked with whole numbers and whole-number fractions. And though philosophically they clearly understood

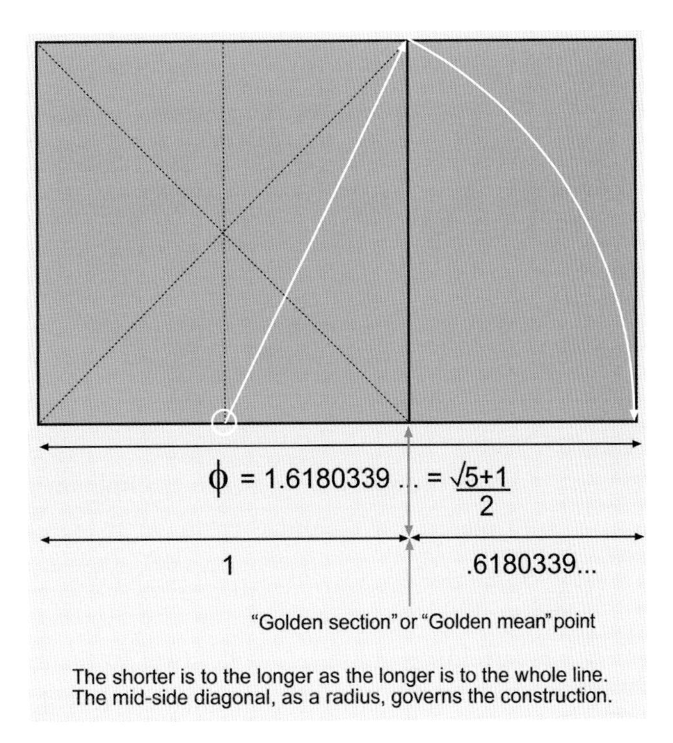

Figure 6.10
Notebook page showing how to divide a line into the golden section. The shorter is to the longer as the longer is to the whole line. The mid-side diagonal, as a radius, governs the construction.

ideals of harmonious proportions, numerically they seem to have had no vocabulary to approach a concept such as ϕ. Some Egyptian iconography suggests that the Egyptians may have known the golden rectangle (a rectangle with sides in the ratio $1:\phi$). It is plausible that they knew how to find the golden section of a line (recall the notebook page in fig. 6.10) since the mid-side diagonal of a square gives the construction for getting it, but they do not seem to have left any wholly convincing visible clue.

For all the amazing skill and sophistication in the planning and construction of the pyramid, the basis of its beauty lies for me in its fundamental geometric and numerical simplicity. The *sheked* of 7:11 means that the ground square of the edifice was geometrically conceived as a square that was 11 by 11. Since the sides are 440 cubits, the conceptual ground grid was a great square of 11 × 40 cubits times 11 × 40 cubits. The more one explores the sand reckoner's square, the more prominent the 11 measure becomes.

NEW MEASURES

"Pyramid power" became a popular New Age interest in the latter decades of the twentieth century, and it is going through an intriguing twenty-first century reincarnation—with a difference. Over the last decade in Russia there have been intensive studies, and indeed commercial applications, of radically redesigned pyramids, now definitely involving ϕ proportions. A scientist friend, Roger Taylor, has told me that experiments there have shown that a pyramid form that fits over two spheres whose diameters are in golden-measure relationship to each other, has been found to have remarkable effects on the crop yields of seeds placed therein; and research by members of the Moscow Academy of Science seems to show a host of other beneficial results.

From the sheer scale of the example on page 72 (fig. 6.11), which is 44 meters high—and there have, to date, been seventeen of these built—one can see that the makers are serious about what they are doing. This structure weighs around 50 tons, being not solid, but built of scaffolding and relatively lightweight materials.

On investigating the geometry I was delighted to find that any two circles with diameters that are in golden proportion to each other (1:1.618 . . .) will sit inside a Starcut square with a side length equal to the sum of their diameters, in such a way that the mid-side diagonals from the bottom corners make exact tangents to the larger lower circle and, of course, meet at the upper center point of the smaller one. So the in-circle of the triangle made by the base and two mid-side diagonals, cuts the center line of the square at a golden-section point (ringed in fig. 6.12). This can be stated as what seems to be a new rider to geometric theorems about the ϕ ratio:

> In an isosceles triangle with the base equal to its altitude the in-circle's circumference cuts the altitude at the golden-section point.

With its simplicity and its relevance, not only to pyramid measures, ancient and modern—and also, as we shall see later, to the classic, Indian, and Chinese canons of proportion and number—one senses that from a very early date the Starcut may have been a working tool both in the development of geometric concepts and in the building of structures. One can see, from figure 6.8 (p. 67), how very easily it could have been used to make a template for checking the slope of the Great Pyramid.

So whereas the Tetraktys—which we will consider in some depth later—was to the Pythagoreans a venerated sacred form in the name of which a vow was made, the Starcut, if known, could have been a more practical device—and, as such, perhaps a more closely kept professional secret. As mentioned earlier, it always brings all its information with it whenever it is drawn, so maybe no one was concerned to engrave it lastingly in rock. It can be carried around in one's head, and sketched as required. After which one could simply let its sands blow away in the wind. It could always be made again tomorrow. But this remains speculation.

Was this diagram actually a device taught in the Egyptian temple schools? Did some ancient Sumerian geometer sit and dream up an entire sexagesimal number system from it? Who can say? There is, to my knowledge, just one tantalizing appearance of a couple of its characteristic mid-side diagonals on a broken cuneiform clay tablet (see fig. 6.13)

Figure 6.11

One of seventeen ϕ-proportion pyramids outside Moscow.

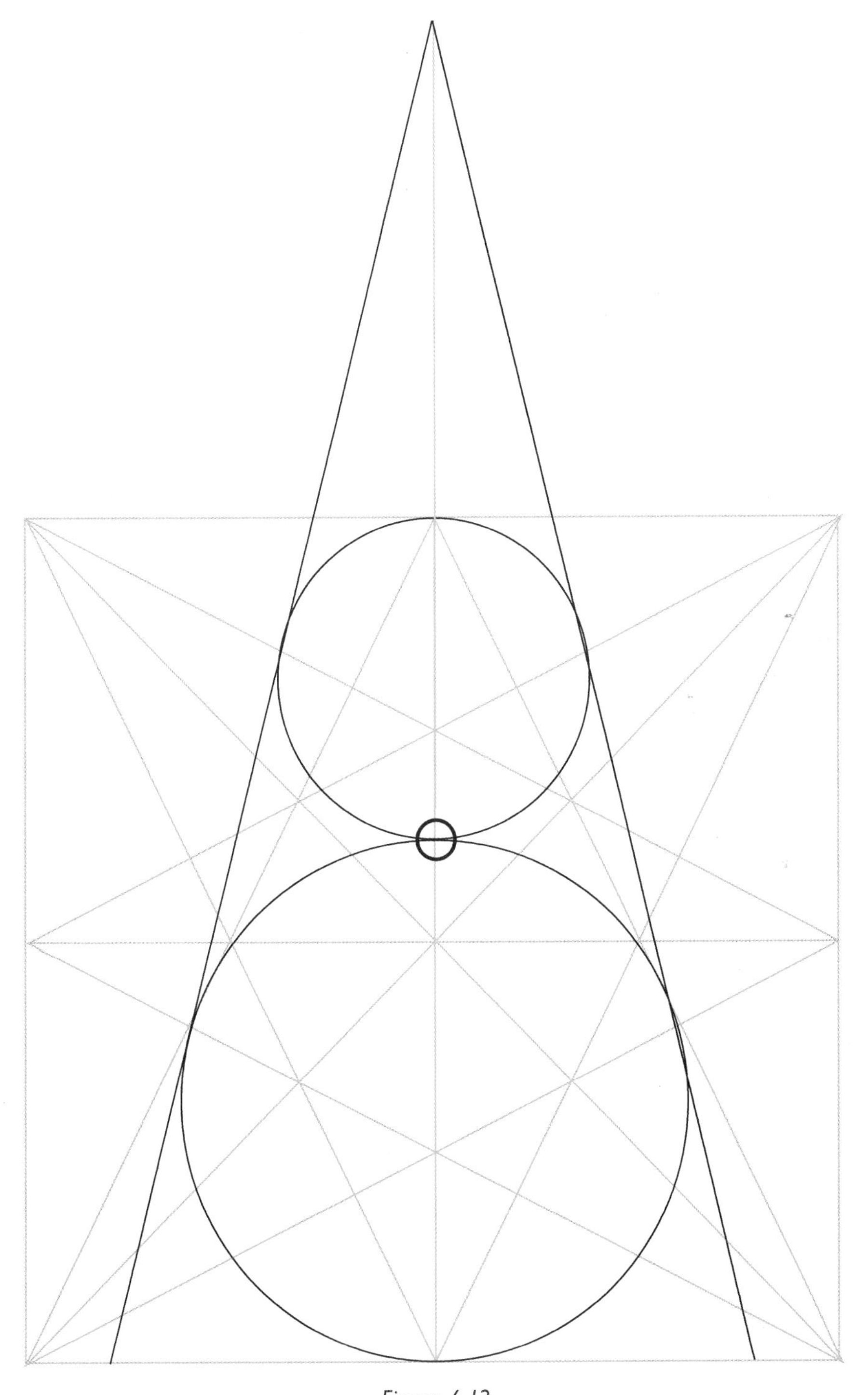

Figure 6.12

In a ϕ-proportioned pyramid the two inscribed circles in the Starcut square divide the center line of the square into the golden section.

dealing with geometry inside squares, however, the whole figure doesn't seem to turn up visibly in a historical context anywhere at all until at least four thousand years after the Great Pyramid was built. But so that the reader should not conclude that the whole thing is a total figment of my imagination, I feel it is time to introduce just such an example.

It was only after nearly twenty years of studying the Starcut as a piece of "sacred geometry," and using it as a design aid in executing glasswork for a mosque, a Hindu shrine, a synagogue, a couple of churches, and a whole series of meditation yantras, that I first met it face-to-face as an explicit figure. The time and place of this occurrence was in the context of Renaissance Spain and, on investigation, the context seemed anything but sacred.

Figure 6.13

Mid-side diagonals on a broken cuneiform clay tablet.

Figure 6.14

This seventeenth-century engraving, somehow simultaneously gruesome and decorous, depicts the final thrust in a duel. But just what is it that the duelists are standing on?

Figure 7.1

The Starcut and the sword art.

7
The Sword Master's Floor Plan

All persons of judgment confess that it would be greatly to be desired, for skill in the exercise of arms, if there were one certain and inviolable measure. . . . Since our subject is the movements which are made by means of the sword, and with the limbs of the body, the most obvious measure would be that found in the proportions of Man himself.

GERARD THIBAULT, *ACADEMY OF THE SWORD*
(TRANS. JOHN MICHAEL GREER)

The first time that I came across a published historical example of the Starcut diagram in pictorial form was in Joy Hancox's book *The Byrom Collection*. The original engraving (fig. 7.1) from Gerard Thibault's *Academy of the Sword* was by Michel Le Blon in the seventeenth century. The diagram provides an underlying lattice upon which are a number of classical figures, notably Zeus who stands in the center in a position suggesting the transfer of power between the above and the below—similar to the position of the Magus in the major arcana of the tarot.

Without going into the detailed associations amid the myths, the astrology, and the four elements (another complex mnemonic!), there are obvious geometric references in the engraving. The central figure, with his vertical staff, can be taken to represent the straight line. The figure in the right upper side, with both hands raised, suggests parallel lines; the figure below this presents the spread form of the pentagon; the figure on the lower left is of the diagonal saltire cross (the Saint Andrew's cross), and above that is a figure denoting the upright meridional cross (for instance, that of Saint George). One clear message therefore is "look at the geometry"; and all the figures, as we can see, are backed by a weblike lattice, which (cf. Allan Brown's rendering, given in fig. 7.2) has a further checkerboard lattice of 8 × 12 subtending it. In figure 7.3 the core of the lattice is shown. In the center, standing on its point, one can see the Starcut diagram.

The figures in the upper corners of the engraving look to me to have been based on Cesare Cesariano's illustrations of Vitruvius's *De Architectura* that appeared in 1521, more than a century before Le Blon (see the example in fig. 7.4).

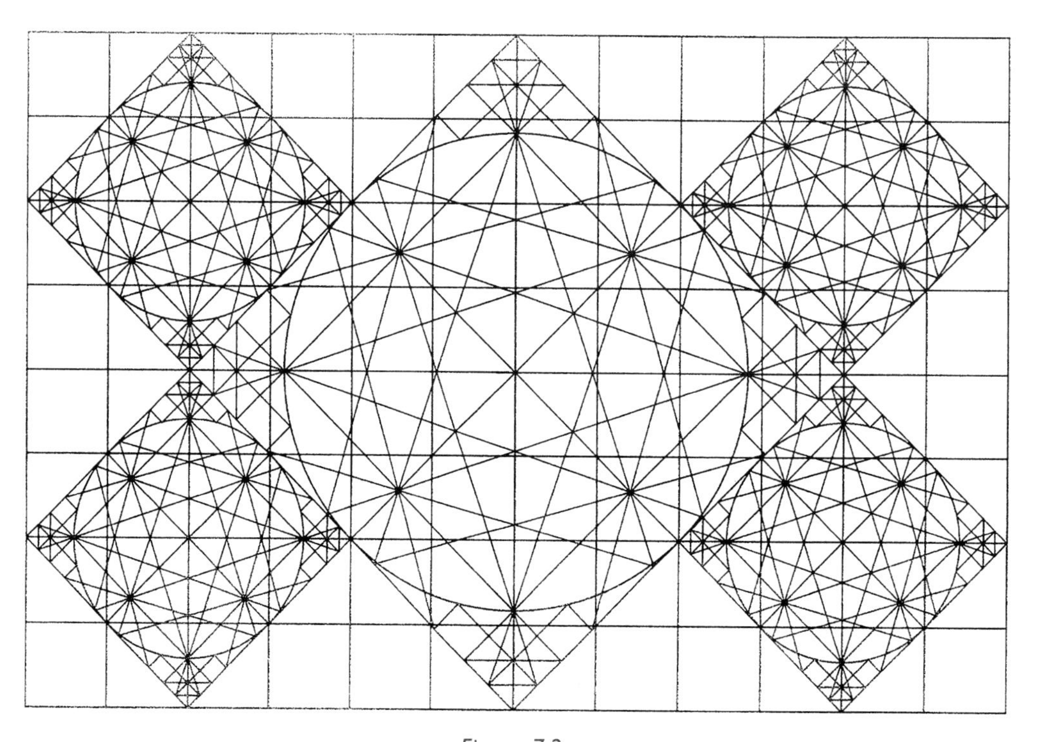

Figure 7.2
Allan Brown's rendering of the underlying lattice.

Figure 7.3
The core of the lattice, revealing the Starcut.

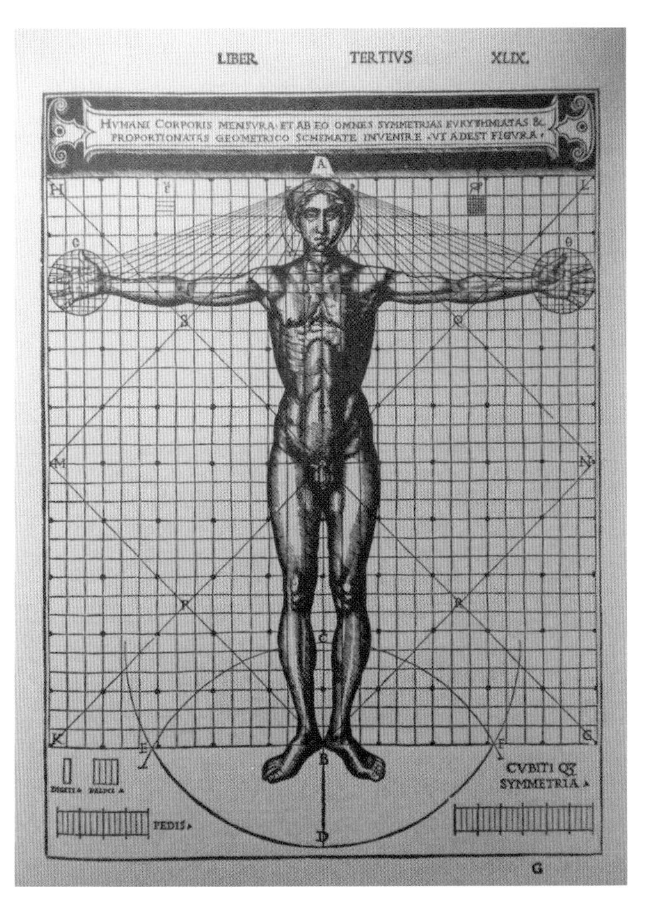

Figure 7.4
Figure from Cesariano's illustration of Vitruvius's *De Architectura.*

In Cesariano's pictures of the body, there is another possible association to do with the grid that he has used. It is 30 × 30—this is a natural division arising from a proportion that the Starcut indicates. We will see that its own internal numerical system suggests that the mid-side diagonals be taken as 30 units long. A 30 × 30 grid therefore becomes natural. This division will be illustrated later in chapter 13 concentrating specifically on the Starcut's inherent number system.

However, in the two hundred or so exquisite illustrations, mostly of a technical geometric nature, that Cesariano executed in his edition of Vitruvius he only once uses the mid-side diagonal that is so characteristic of the Starcut. I have highlighted it, for emphasis, in the lower left corner of Cesariano's page as shown in figure 7.5. Figure 7.6 elucidates the geometry he uses.

In figure 7.5, tucked away at the corner of a page, Cesariano shows an elegant construction where the corners of squares bounding a half-square right-angled triangle are connected to the corners of that triangle, which are also the corners of the two upper squares. I have ringed these points in figure 7.6. He then draws two further lines, from the bottom corners of the big square up to the points (which I have boxed) where his earlier lines intersected. These latter lines perfectly divide the top of the big lower square into two. The figure begins to reveal the Starcut, hanging from a point within the half-square.

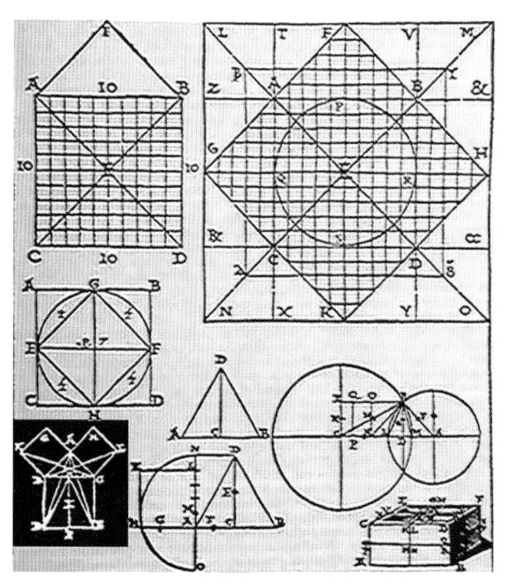
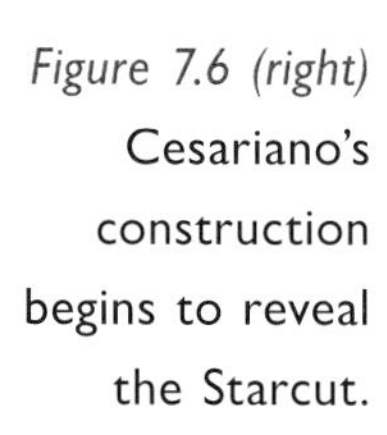

Figure 7.5 (left) Cesariano's construction highlighted in the lower left.

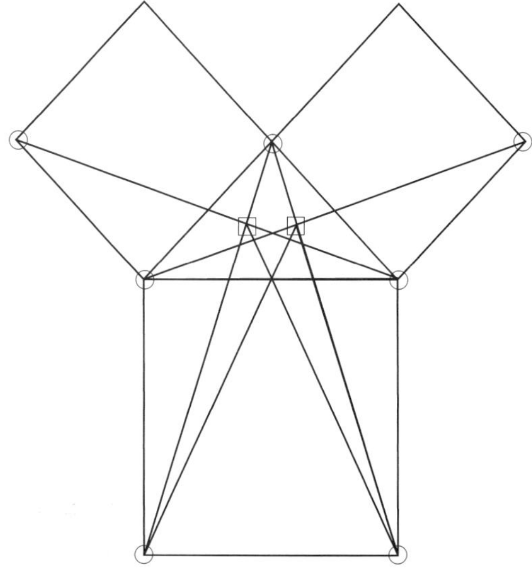

Figure 7.6 (right) Cesariano's construction begins to reveal the Starcut.

Around the same time as Cesariano, the works of Cornelius Agrippa of Nettesheim carry similar images of the body. The human form and its harmonious measures had become a topic of scholarly, and indeed magical, interest (fig. 7.7).

We will return to Agrippa, in another connection, later in this study (see chapter 25). Meanwhile there was, quoted in Joy Hancox's book, another image involving the lattice under discussion. Upon it a figure, this time a female, has her left rather than her right hand raised. She is backed by bodies in positions similar to those in the corners of the Zeus picture. As in figure 7.1, almost all people pictured are carrying swords.

Joy Hancox's researches established that the geometric lattice involved was the floor plan used in the training practices of a rapier

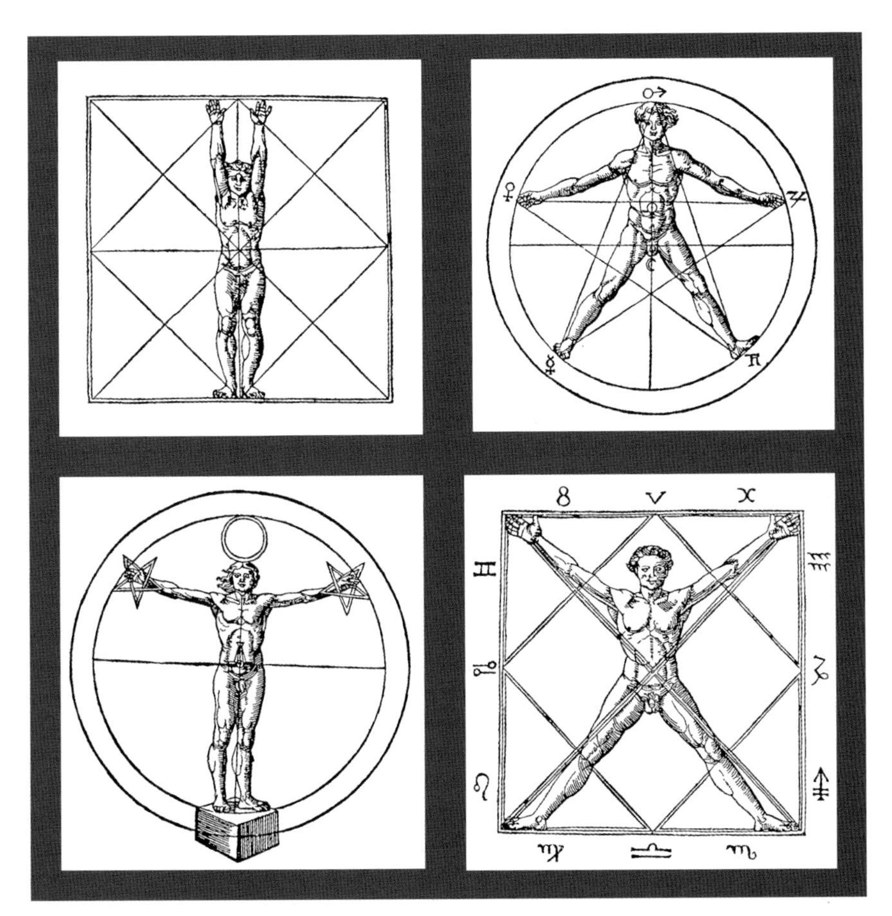

Figure 7.7

The human form in the works of Cornelius Agrippa.

fencing style called *la destreza* ("the skill"). This was, in its time, the most formidable sword form of Western martial arts. Known often as "the Spanish circle," it dominated throughout the centuries of rapier dueling from the late 1500s onward. The technique itself originated with Jerónimo de Carranza in 1569. Figures 7.1 as well as 7.8–7.10) are all from *L'Académie de L'Épee* written about *la destreza* in the early 1600s by Gerard Thibault. It is said to have been the only esoteric martial art ever developed in the West. On the frontispiece, Thibault's book claims to demonstrate:

Figure 7.8

Illustration by Michel Le Blon, from *L'Académie de L'Épee* written about *la destreza* in the early 1600s by Gerard Thibault. Note the geometry at the center of the image.

according to mathematical rules, on the basis of a Mysterious Circle, the theory and practice of true and, until this time, unknown secrets in the handling of weapons both on foot and on horseback.

The illustrations were all done by the seventeenth-century engraver and designer Michel Le Blon. In them the geometry is always present.

After two decades investigating the Starcut with square and compasses, and always assuming it to be a "sacred" figure, I was intrigued that its first explicit occurrence in the historical record should be a martial application.

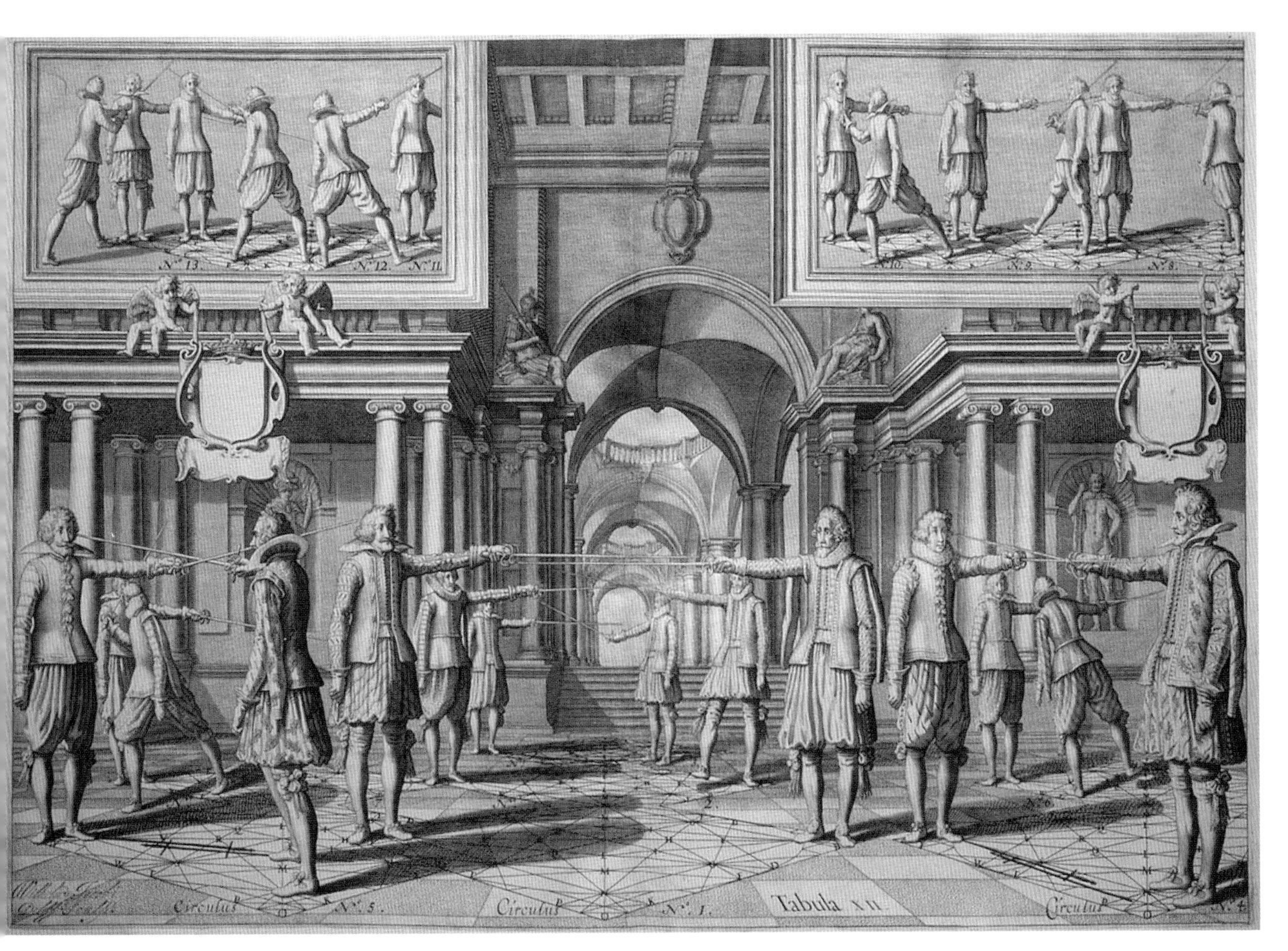

Figure 7.9

Illustration by Michel Le Blon, from *L'Académie de L'Épee* written about *la destreza* in the early 1600s by Gerard Thibault. Note the geometry on the practice floor.

Carranza, the founder of *la destreza,* had himself written a book in which he presents his reasons for his geometric approach:

> You know that mathematics strips bare the forms and figures and numbers of the material, in which no falsehood is admitted, because it does not dissimulate, either affirming or denying, because it considers things simply and not all together. And it has this privilege more than the other arts in that it declares its intentions with most true demonstrations, for which reason the ancients guided those things they called true and certain arts with this reasoning, and that any truth one has in human affairs lies in mathematics because its teaching is very pointed (that is to say, simple); it guides the sciences, whether moral or natural, by the most direct route . . . as is seen in the beginning of the Sextant of Euclid, from whence is born the foundation by which the astronomers verify their calculations which are found most copiously in the Almagest, and whose esteem and importance will be seen in the inscription that was in the School of Plato, which said that no one who was ignorant of mathematics could enter inside. This is the reason it seems to me a more certain thing to order la destreza under mathematics rather than any of the other sciences.

The detail in figure 7.10 is hardly readable, but the general schema behind Thibault's image is as follows. The upper-left tilted square is keyed to an elaborate description, in the text, of how to draw the floor plan such that it is in proportion to the practitioner's body, his movements, and his sword, whose length is also specified. The upper-right tilted panel notes key points on the skeleton. The lower left and right panels show body proportions, front and back, both upright and inverted, to highlight bodily symmetries mentioned in the text. I have inlaid a simplified version of the original central panel which, in the book itself, is festooned with detailed markings giving twenty-five or so exact points of anatomy to be mastered by the practitioner to facilitate his art.

The "Zeus" figure can now be seen to be the upright swordsman's body position. This position is particularly appropriate to this rapier school, which was known for the formal uprightness of posture of the practitioner. The English sword master George Silver, writing in the 1690s, comments on this:

> The Spaniard is now thought to be a better man with his rapier than is the Italian, Frenchman, high Almaine or any other country man whatsoever. . . . This is the manner of the Spanish fight. They stand as brave as they can with their bodies straight upright, narrow spaced, with their feet continually moving, as if they were in a

Figure 7.10

Illustration by Michel Le Blon, from *L'Académie de L'Épee* written about *la destreza* in the early 1600s by Gerard Thibault. The geometry is seen on the floor and throughout the image.

dance, holding forth their arms and rapiers very straight against the face or bodies of their enemies.

The lines of the lattice become lines of body measures, foot movement, advance, and thrust. Foot positions are marked, somewhat as the foot positions occasionally appear on instructional diagrams for the Chinese art of tai chi chuan. The diagram was considered to define the proportions of the body, in a similar manner to the "Vitruvian Man" of Leonardo, Cesariano, or Agrippa, but with a much higher degree of accuracy. A chapter of Thibault's book is devoted to a favorable comparison of this body map to that which had been developed by Albrecht Dürer. Additionally in the Spanish circle usage, however, the map is concerned to define the positions of the body's most vulnerable organs. As Carranza, the originator of the art had written: "The Body of Man is the first foundation of the Skill."

Carranza's own book had been full of classical allusions, with some of it being written in a style somewhat like the question-and-answer style of Plato's Socratic dialogues. Certainly Carranza considered *la destreza* a philosophic art appropriate to a man of honor and virtue. Though hardly "nonaggressive" in the sense that tai chi chuan is said to be, *la destreza* seems to have been essentially defensive, if only from caution. Carranza advised the exponent of the skill to be wary: "He who knows most, doubts most."

A huge corpus of literature about *la destreza* has grown over the centuries, and modern students of the art of rapier still discuss and use it. I understand that translators are working on an English language edition of Carranza's "Philosophy of Arms"—a big sprawling work that sought to include references to every field of Renaissance study; and the *la destreza* lattice has been used as the logo of the American Association for Historical Fencing. It is not in all respects identical to the Starcut diagram only because some extra lines cross the figure connecting it to the complex borders and corners that are, it seems, wholly concerned with its application to fencing.

In view of what I already knew of the center of this diagram I could not believe that it had originated with Carranza. So I looked for it elsewhere in the Renaissance arts. I knew that the designer Michel Le Blon, who must have known the diagram intimately, had ancestral family connections with the earlier artist, geometer, and cartographer Théodore de Bry who seems, in turn, to have known Leonardo. The mid-side diagonal aspect of the diagram was indeed to be found, applied to the profile of the head, in Leonardo's illustrations of Luca Pacioli's *De Divina Proportione.* This form is also to be found, but perhaps without further significance, in Dürer (see fig. 7.11); but it does not seem to appear elsewhere in his or Leonardo's work.

There was, however, another artist, younger than Leonardo, in whose work I had already found strong traces of the classical geometric tradition: Raffaele Sanzio, famous to us simply as Raphael.

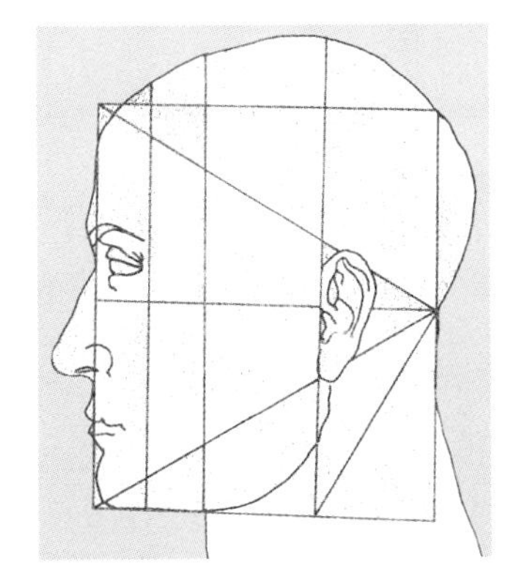

Figure 7.11
The mid-side diagonal aspect of the diagram was indeed found in this Albrect Dürer study of the human head.

Figure 7.12
The words on the pediment are: "The Book of Jerónimo de Carranza, born in Seville. A treatise on the Philosophy of Arms and the True Skill of Christian Attack and Defense."

8
The Hidden Geometry of the "Divine Raphael"

It was the author Trevor Ravenscroft who had first told me that Raphael "was an initiate." When stripped of its occult associations, this translates to mean that Raphael had been educated in the Platonism that was the main driver of Renaissance thought and art. This was naturally the case since his father and first mentor Giovanni Sanzio seems to have been the general arts consultant (some say he was the court painter) in Urbino. The court of Urbino had been established by the Duke of Montefeltro who, though a mercenary warrior, nonetheless aspired to the ideals of Christian Platonism and that his court should be "the dwelling place of the muses." The famous handbook of idealized manners *The Courtier* by Baldassare Castiglione was written in Urbino. Many leading names in painting, sculpture, mathematics, architecture, and engineering had frequented the court. Among them were Paolo Uccello, the great pioneer of perspective in painting; Piero della Francesca, whose beautiful works are also object lessons in geometric and perspectival cohesion; Luca Pacioli, the mathematician whose book on proportion was illustrated by his friend Leonardo da Vinci; and Roberto Valturio, whose seminal book on the military arts and engineering was illustrated with technical and exquisite woodcuts that greatly influenced Leonardo's work in that field.

Opposite: Figure 8.1 A self-portrait by Raphael.

Raphael grew up in this cultural environment, and was apprenticed at the studios of Perugino and Timoteo Viti, after whom a street in Urbino is still named—with a plaque proclaiming Viti as the tutor of the "Divine Raphael." One can see, in Viti's painting (see fig. 8.2) the characteristic interest in perspective and, in the central figures, the relaxation of posture and gesture within a formal setting that were to be present throughout much of Raphael's work.

There is some fundamental geometry in Raphael. I first noticed it in his last masterpiece *The Transfiguration* (fig. 8.3), that was completed by one of his many followers, Giulio Romano, after the young master's untimely death aged only thirty-seven. The painting illustrates two New Testament stories that occur in the gospels of Mark, Matthew, and Luke. They tell of Jesus going with his disciples Peter, James, and John up a mountain—traditionally Mount Tabor. There he converses with Elijah, Judaism's archetypal prophet, and Moses, its archetypal lawgiver.

Figure 8.2
Timoteo Viti's painting of Thomas Becket and Saint Martin, with Archbishop Arrivabene and the Duke of Urbino.

Thereafter Jesus appears transformed in a blazing light; whereupon a "voice from heaven" proclaims him as "my beloved son."

As a stained glass designer used to working to architect's specifications I habitually notice the overall proportions of a rectangular work of art. The fact that Raphael's picture perfectly fitted the boundary rectangle of two circles interlocking as a vesica did not seem to have been an accident. As a ratio, giving its short side a length of 2 (the two radii of the circle that is the width of the painting), it is 2:3. This ratio is the "perfect fifth" harmony in music, for instance, the notes C and G played together, or in sequence, as in the opening notes of "The Last Post." It is the first natural consonance after the octave and is basic for the derivation of the chromatic twelve-note scale, which ascends in a spiraling ratio until it completes itself by finally repeating a note that has occurred before—though it is now seven octaves higher. This whole series is called

Figure 8.3
Raphael's
The Transfiguration.

"the cycle of fifths," the final note of which is not quite true—but that, too, will be explored later in chapter 19. The simple 2:3 ratio, root of all this, is as harmonious to the eye as to the ear or to the mind.

Raphael's image of Christ is in a form slightly echoing that of crucifixion, and is simultaneously a preecho of Resurrection. He is pictured up beyond the cross form of the underlying geometry (see fig. 8.5), the central line of which comes down through his heart and his second toe(!). Thereafter that line is reflected in the direction of gaze of all the characters in the painting: the prophet and the lawgiver and the two disciples who are looking up at the vision. In the lower dark zone of the painting, the gaze of all the characters again reflect across that midline; except for a single individual, the second finger of whose left hand, like the second toe of Christ's right foot, is right on this line. It's as though his interception of it allows him to have his face across it, thus breaking any over-formality in the composition.

All three Synoptic writers follow the second story of the transfiguration with that of Jesus, having returned from the mountain, healing this boy whom the apostles have failed to cure. The child is afflicted with a convulsive and vocal complaint. Mark and Matthew report he is dumb, Luke (a doctor) says that the boy is prone to "crying out." Both versions agree that his vocal system is a problem. That's worth noting for later.

That the shape of the painting should be the rectangle of the circles that produce the vesica piscis, the bladder of a fish, is appropriate to the key figure of the Christ, known soon after his own time by the Greek word for fish, *ichthys,* derived from the beginnings of the words Iesous Christos Theou Yios Soter (Jesus Christ Son of God, Savior).

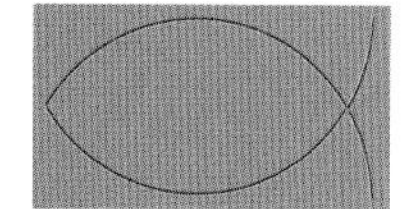

Figure 8.4 The vesica symbol as a fish, representing Christ.

We know how two Christians could recognize each other during the Roman persecution. One of them would casually trace, with his or her foot, a curved line in the sandy ground; if the other replied with a line that came together with the one already drawn, into the shape of a fish, they knew they were of the same faith (fig. 8.4). (Nowadays we are familiar with this symbol telling us that the car in front is being driven by a Christian.) So the vesica seems very fitting to a painting that pictured this teaching about Christ as saving healer and divine son.

Figure 8.5
The two circles forming a vesica in *The Transfiguration*.

On developing the geometry into the classic esoteric division of the semicircle and triangle of overarching spirit above the square of materiality (fig. 8.6), one can see how the painting is divided into two connected narrative zones. In the top half Jesus is transfigured, bathed in light, and attended by Elias and Moses. Immediately below Jesus are

Figure 8.6
Geometric division of *The Transfiguration.*

the disciples Peter, James, and John, two of whom are shading their eyes to peer toward the light, just managing to cross the line into the zone of vision. Over to one side are two figures, the front one of whom is—maybe—the patron of the picture, with the one in the shadows being the artist perhaps?

In the bottom half of the painting, we find a busy scene concerning the healing of the boy with convulsions. However, we don't actually see the healing but the disciples' failure to heal. In the Gospel story it is only Jesus who can do this, and he comments that it takes a lot of "prayer and fasting" to heal certain things. But is the healing itself anywhere in the painting? It's certainly not on the surface.

Figure 8.7
The five-pointed star overdrawn on Raphael's painting.

Purely as an experiment, however, I decided to draw a regular five-pointed star to fit as far as possible into the most obvious place, with its point crowning the Christ figure and its width spanning the whole canvas (fig. 8.7).

Without any cheating, the radiance of the pentangle was directly touching the center of the boy's condition. The Pythagoreans had considered the pentagram to be the figure that best represented the perfect human—which for Christians equated to Christ. It is imbued with the geometry of the golden mean, which also pervades living systems—our own human body very much included. By natural analogy the pentagram was thus the figure for healing.

It is difficult to imagine that all the above happened by pure chance. There is more in this painting, but what we have seen seems to affirm that Raphael's masterpiece of representation is draped on a lattice of geometric meaning married to the painting's theme and narrative—"sacred" geometry indeed.

CAUSARUM COGNITIO

This famous fresco painted by Raphael during 1510 has since some time in the seventeenth century been called *The School of Athens* (fig. 8.8), but the name by which the artist himself knew it is given in the ceiling medallion (fig. 8.9) that is above it in the papal apartment

Figure 8.8
Raphael's *The School of Athens.*

Figure 8.9
The ceiling medallion naming the fresco as *Causarum Cognitio.*

Figure 8.10
The celestial architecture representing the "harmonious beauty" of the thought-built cosmos, with Apollo on the left and Athena on the right.

known as the Stanza della Segnatura (the Room of the Signature). The name given there is *Causarum Cognitio*—"Knowledge of Causes"—a phrase coined by Cicero. The medallion directly hints much about the fresco beneath it. The female figure of philosophy, dressed in the colors of the elements, holds two books, one vertically, which is titled *Morals,* and one horizontally, which is titled *Nature*. This surely comments on the two central figures of the fresco, where Plato points upward to the heavens and Aristotle gestures to the earth below.

Raphael's own title adds an extra depth to the interpretation of the entire work. The figures in it are not just a dramatis personae of classical philosophical history; they represent various ideas as to what are the causal fundamentals of the real world as grasped by the world of thought. For Raphael and his educated contemporaries, the Athenian Lyceum was, above all, the Temple of Thought. It was Greek thought that had brought fresh light to dogma-bound European culture. The liberal arts were indeed a liberation. Not for nothing are the heads of the two great philosophers, Plato and Aristotle, depicted against the open sky rather than backed by an altar screen (fig. 8.11).

Clearly Raphael is echoing the architecture that was, at the very time he was painting this fresco, being built in Saint Peter's Basilica, "next door" to the Vatican chamber where he was working; although the environment in his painting has classical rather than ecclesial references.

Above the figures Raphael presents the thought-built *cosmos*. The word *cosmos* means something like "harmonious beauty." The left of the picture is beneath the patronage of Apollo, the god of both the Pythagorean arts and of prophecy. He was considered to be the one who preserved the heart of Dionysus—the divine aspect of humanity expressed via the studies that surround our theme. The right of the fresco is under the patronage of Athena, goddess of clear reason, who leapt fully armed from the brow of Zeus and whose discriminative understanding and reflective power were a match for the confusion of the riddling Sphinx and the mind-paralyzing fear caused by the Gorgon. Beyond them Raphael gives the imagined architecture a celestial scale; one can see a glimpse of a huge central dome going into the blue sky above and beyond the living figures. The temple's tall supporting walls are niched with hidden gods. It is the logic of this vast, clearly lit space that becomes the overarching concept within which the newly educated Renaissance mind celebrates the culture that has inspired it. Plato and Aristotle, archetypal personifications of Athenian philosophy, stroll in peripatetic discussion as did the Athenians beneath the porticos of the stoa on the agora; but here the walkway is more like a stage.

Plato is pointing to heaven and carries the book of his *Timaeus,* the dialogue where the creation is described as a geometric and elemental process of the Divine Mind in action (fig. 8.11). Plato taught too that union with that Divine Mind was accessible through a philosophical path of ascent. The causal ideas for him are themselves the vehicle by which the philosophic mind is guided on this path.

At Plato's left shoulder is Aristotle. His hand is palm down, fingers extended, indicating the earth. He carries the *Nicomachean Ethics,* his own great moral dissertation. He is here the patron figure of earth-bound reason and virtue—with its consequent happiness that he called *eudaimonia,* meaning "good spirit." Aristotle also represents the natural sciences and the human arts. Socrates is in green, on the same level, teaching a small group off to the left of Plato, counting off the points of his logic on the fingers of his hand (fig. 8.12).

The front apron of this theatre of thought features Pythagoras, closest to the viewer in the left foreground. His cosmology is associated

Figure 8.11 (left) Plato and Aristotle.

Figure 8.12 (right) Socrates teaching.

with the Tetraktys, with numbers and with the ratios of musical harmony. These appear on the board that is being held in front of him.

The figure in white, looking out of the picture (fig. 8.13) is said originally to have been Hypatia, mathematician and geometer, the leading Platonist teacher in Alexandria around 400 CE. The story has it that Raphael was warned by a bishop against including her, and this figure is what he painted instead. The figure's gaze contains a challenge; perhaps Raphael was reminding Julius II, his papal patron, that Hypatia had been viciously flayed to death by a Christian mob with, it was thought, the connivance of Cyril, the "pope of Alexandria," later canonized as a saint.

Heraclitus, with the intense face of Michelangelo (Raphael and Michelangelo did not get on well), leans against a cubic stone and muses on the celestial fire that for him caused both creation and the passing nature of things (see fig. 8.13).

Zoroaster (fig. 8.14), teacher of Magian star wisdom, stands in a group to the right with the celestial globe in his hand. Raphael himself gazes out of the fresco from this group. And bending down, with the face of the architect Bramante, in the right foreground, is either Euclid, or some say Archimedes, with a pair of dividers expounding geometry to a group of students.

Figure 8.13
Pythagoras's board, with Hypatia (in white) and Heraclitus (seated right).

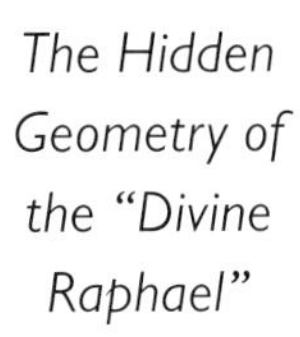

Figure 8.14
The group around Zoroaster, and in the foreground the geometer Euclid, or Archimedes, holds the dividers.

Sprawled with untidy freedom on the steps in the center is Diogenes (fig. 8.15) the feral cynic who, while living naked and with no possessions in his barrel, was asked by Alexander if he wished for any favor and requested only that the conqueror should stand aside because he was blocking the sun. Raphael encodes Diogenes's naked detachment by garbing him in the color of the sky.

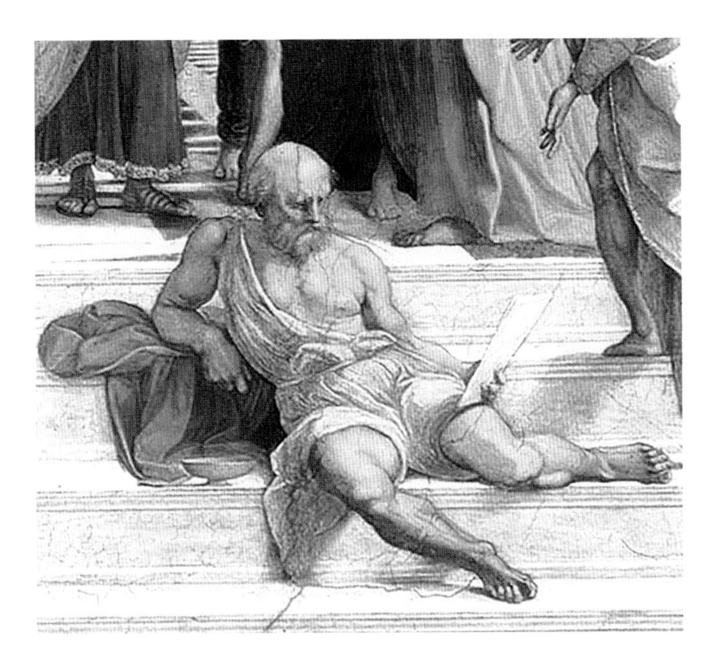

Figure 8.15
Diogenes.

I have only skimmed the allusions in the painting, of which there are many more, to illustrate that Raphael was thoroughly versed in the classical tradition. However, he enshrines his characters in a classical-seeming symmetry that is, in fact, pure Renaissance.

THE CENTRAL STARCUT

So coming now to the geometry of the picture, the first general form is obviously the semicircle. It is the structural shape of the niche in which the fresco sits. It is reflected into the picture as the half-barrel ceiling vaults that are the undersides of the triumph-style arches supporting the roof and that huge, mainly unseen, dome. Semicircular architectural features were very much a hallmark of fifteenth- and early sixteenth-century painting, but here this formalism is particularly appropriate since Pythagoras, the foreground figure seminal to the whole tradition,

Figure 8.16
The entire painting within the semicircle.

Figure 8.17

Key governing points of the Starcut have been ringed.

called his school "the Semicircle." This was perhaps out of respect for Thales who was said to have been Pythagoras's first teacher and who was the original inventor of fundamental theorems concerning angles within circles; notably the fact that all angles spanning semicircles are right angles.

The diameter of the semicircle marks the upper-floor level where Plato and Aristotle hold court. The geometric "plumb line" of the picture goes down from the center top and passes along the edge of the cube on which Heraclitus leans in thought (see fig. 8.17). The cube traditionally represents pure reasoning and perfect mind, based in three-dimensional reality. Heraclitus's gaze leads us to the perspective indicated by the tiling on the floor; and it is via this fundamental perspective of the entire painting that we find the Starcut diagram, which emerges as the primary "causative" principle of the entire architectural construction.

The cornices receding into the arched vault above dictate the lines of perspective to their disappearing point, which lies exactly on the plumb line between Aristotle and Plato. The points of the foremost cornices dictate the square in which the two lines of perspective are shown to be the mid-side diagonals. The upper center of that square marks the far edge of the underside of the arched vaulting. In figure 8.17 I have ringed that point and also the cornice points of the second archway that lie on the transverse midpoints in the lattice. The perspective lines, exactly along the long cornice edges throughout, continue downward into the foreground to guide the aspect of the tiling. I have marked their extensions all the way to the bulls' heads at the apparent base of the fresco.

Only the heads of the two central philosophers are within the downward triangle of the mid-side diagonals. The reader will see from figure 8.17 that the Starcut and its extensions govern many aspects of the fresco's overall balance. Again, as with *The Transfiguration,* Raphael's artistry integrates his subject. I do not think it fanciful to note that the diagonals of the Starcut continue to touch the head of Pythagoras on the left and Euclid's drawing arm on the right. These were great seminal figures of the classical geometric tradition, and I believe the geometry references them in this way. It is worth noting too that the transverse line through the Starcut nodes closest to the heads of Plato and Aristotle governs the general height of the head and eye lines of all the figures in the upper scene. The circumcircle of the Starcut square is found to define exactly where Plato's and Aristotle's feet are placed.

Although there are small deviations from perfect geometric accuracy, inevitable in the execution and rendering of a large finished picture, it seems beyond doubt that Raphael used a central Starcut square as an exact governor of the picture's perspective and a more general guide to other aspects.

The theme here, we remember, is not lightly chosen. The picture is in Pope Julius's apartments, the invitation to decorate which represented the pinnacle of what a painter of Raphael's time could hope to achieve. On another wall he was to depict the angel effecting the escape of the apostle Peter from prison—a clear bow in the direction of the pope, considered to be Peter's apostolic successor. Here in *The Knowledge of Causes* he pictures not just any classical theme but the very core of what was sweeping educated Europe. The Athenian Lyceum was *the* school for all Renaissance courts including that of the ruler of Christendom. If the Starcut was the central geometric determinant for Raphael's formal depiction of classical philosophy, it suggests that it was a known authoritative device of earlier provenance, and that someone may have passed it on to him as such. I suspected that the source might be the painter and architect Donato Bramante, the man whose face was used for the geometer (Euclid or Archimedes) in the right foreground of the fresco. Bramante, a generation older than Raphael, was also from Urbino and was known for his mastery of the Gothic and Byzantine and the fusion of these with the classical that characterized the Renaissance. Bramante had known Mantegna and Piero della Francesca, both renowned for work with perspective and both influential in the court where Raphael grew up. It had been Bramante who had first recommended Raphael to Pope Julius; and, further yet, Bramante had drafted the first proposed ground plan for the basilica whose style was echoed in Raphael's fresco. It was not until I read Joscelyn Godwin's *The Pagan Dream of the Renaissance* that I learned of a reference in Vasari's biography of the artists of his time. He had been in Rome a couple of decades after Raphael died and had made a special study of Raphael's work; he said that Bramante should be given credit for the architectural aspect of the *Causarum Cognitio*. I looked for something of Bramante's that was both square and contained sufficient information to give some clues.

His original ground plan for Saint Peter's seemed a likely match, but the original of this is faded and not reproducible. Over the centuries, copies have been made. Inevitably they have ended up very distorted.

Figure 8.18

Bramante's original ground plan for Saint Peter's, Rome.

Hence I apologize for the blurred version in figure 8.18. It is the survivor of centuries of rough handling plus my attempts to pull it back into a true-ish square. I do believe that it was a Starcut that Bramante used underlying his original basilica plan. There are many correspondences with the pattern; the most persuasive being the pairs of ringed dots adjacent to the corner octagons. They happen on lines, the mid-side diagonals, that are an exclusive characteristic of the diagram.

The next chapter is an interlude about Delphi and Apollo. As patron of the muses and the arts of harmony and proportion, Apollo was said to be the only "god" venerated by the Pythagoreans. He embodies a union of qualitative and quantitative thought, of magic and logic.

Figure 8.19

Synchronicity: This is a true story. In 2002, my wife and I were visiting South Africa, staying in a beach house in Betty's Bay. One morning we noticed a house that, from its distant profile, looked to have interesting proportions with an attractive finish. We had never before taken a detour to look at a particular building, but we drove a few hundred yards down the side road for a closer look. When we came alongside the front of the house, we saw, astonishingly, that the frosted-glass windowpane above the front door, completely invisible from where we had started our detour, was decorated with the Starcut. Herewith the holiday snap.

Figure 9.1
The Tholos at Delphi.

9
Guardian Apollo

Who walks thinks; who thinks
Gets all he deserves.

John Esam, "The Yellow Ancestor,"
from *Orpheus, Eurydice: Songs Late and Early*

While the liberal arts flourished anew during the Renaissance, they had been identified in the medieval period that was soaked in some of Aristotle's work after the Christian armies had overrun Toledo, and the Arabian translations of his writings had become more available to European scholars. Versions of Plato had been around centuries earlier, largely as translated into Christendom via Boethius and Augustine of Hippo.

So far as I know, the actual term "liberal arts" first spread in the eleventh- and twelfth-century centers of privileged and ecclesiastically dominated learning—the monasteries and the universities. The elite aspect of all this was somewhat moderated by the fact that an ordinary monk, if very bright, just might in the end become an abbot—the equal of a lord or a bishop. Educationally the monasteries introduced an early glimmer of what we would now call social mobility, through a distinction, according to aptitude, between "servile" and "liberal" arts.

Figure 9.2
Reconstruction of the stoa of Attalos on the agora in Athens.

In classical Greece these liberal, as opposed to servile, arts had been the interest of the philosophers known as Peripatetics ("strollers") who ambled and discussed the verities under the colonnades of the stoa. Due to our fascination with the "pythonic" oracle at Delphi it is often forgotten that the Athenians established a stoa there too; a place to stroll in the warm shade and practice their discipline of learning-by-talking.

The Socratic dialogues are perfect examples of this art. Some schools actually specified the proportion of a day that the pupil should spend talking. Apollo's Delphic shrine, therefore, was not revered from an exclusively divinatory point of view. It was the center of intelligence and the arts both in legend and in fact.

In addition to the temple, oracle, and stoa, there was a theatre and a tholos (shown in fig. 9.1). The theatre would have, first and foremost,

Figure 9.3
The remains of the Athenian stoa in Delphi.

housed the dramas associated with Apollo as the patron god: his first shrine, his battle with the Python, his bringing of the Delphic culture, and his own cleansing, through exile and servitude, from the guilt that he incurred by killing the Python.

There is uncertainty about what purpose the tholos served. Its shape—a raised colonnaded circle—probably indicates some platform-in-the-round function. It was decorated with a frieze of the battle of centaurs and amazons. Presumably it was the place for bardic narration, for music, for poetry, for proclamations and public announcements. As such it would be a necessary part of the facilities for the eight-yearly bardic "Pythian games" involving poetry contests and music. These games came to include athletic and martial games leading to the building and rebuilding of a stadium above the older site.

Lower on the hillside, Delphi also had treasuries where were stored the "votive offerings" of its wide portfolio of powerful sponsors: Arcadians, Corinthians, Athenians, and significant others. The stores were "votive" because those sponsors had "vowed" to protect the place, and proved their word by storing their own riches there.

The legends about the site had it first sacred to Ge (Gaia) and her son Python. Apollo had come, reinforced with what old texts called the magical "dolphin" assistance of the Cretans. He had made the place "Delphoi" after overcoming Python and casting him into the cleft by the Castalian Spring (see fig. 9.4).

Figure 9.4
The cleft above the Castalian Spring. Recent scientific studies have confirmed that traces of ethylene still sometimes seep up from beneath the ground here. Ethylene can be psychoactive; in high concentrations it can be lethal.

Apollo had then entombed the nymph Castalia who had maliciously lured him into that "dragon infested" region in the first place. In other tales, he transformed her into the spring itself. It seems that these ideas of prophecy, of the female beneath the ground, and of the dragon/python cast into the deep cleft, became somehow blended in the practice of the Pythian oracle sitting in an altered state of consciousness, on a tetrahedral tripod above the ethylene-emitting crevasse, from where she pronounced the oracle whose interpretation was the business of Delphi's priests.

Other figurative sources show Apollo in another legendary confrontation, this time with Hercules. Again it is for control of the prophetic tripod.

Apollo's furious aspect is by no means only featured in Delphic traditions. The *Iliad* opens with him firing burning arrows of fever into the Achaeans in their ships, having already shot down their animals. The god has been angered by an insult to Chryses his priest by the sons of Atreus. In another tale we find him flaying Marsyas the flute player over whom he has triumphed in a musical contest.

Yet it is said that Apollo was the only god venerated by the Pythagoreans. He fathered Asclepius, the hero of healing, and was the god of the muses and the lyre. The late Anne Macaulay, a historian

Figure 9.5
Apollo, god of the muses and the lyre.

of the alphabet and of Pythagorean studies and music, traced the tuning of the top four strings of a modern guitar to Apollonian roots. (The pitches, bottom upward, are D, G, B, and E.) She invoked the belief that Hyperborean Britain was Apollo's birthplace where his two constellations Lyra (the Lyre) and Cygnus (the Swan) never set in the night sky; whereas the god's celestial symbols disappear from the skies of Delphi and Delos, his traditional birthplace, for some months each year. During this time he was said to have gone north. Anne Macaulay made a case for the British druids having the same tuning as did Apollo's lyre, a tuning incorporating both the major and the minor triads of harmony. We shall come back to these.

In the lore associated with the Mysteries of Eleusis, Apollo was the guardian of the heart of Dionysus who had been torn apart by the Titans. Dionysus was "god in human form." The Titans—earthbound forces—had incorporated the divine into their flesh by eating what they could of Dionysus. It was taboo for Apollo to kill those who had something of the god in their bodies. But what he could do was to carry the sacred heart to safety becoming, as it were, custodian of the heart's divine thoughts of the One, the Good, the Beautiful, and the True. The arts dedicated to Apollo were considered to embody the universal disciplines flowing from that higher spirit.

In these terms the "liberal" arts really are free and equally open to anyone—and can be called "Apollonian" arts, coming from a knowledge of measures that are true, beautiful, and good for us, body, brain, and soul. They survive the fashions of our Titanic commercialized society by transcending them as archetypal simplicities of form, number, and harmony expressing integration of the One and the Many. They are perennial—unanimous.

Opposite: Figure 9.6
Two coins commemorating the oracular tripod. On the *top* we see serpentine symbols within the image; on the *bottom* is pictured the battle of Apollo and the Python for control of the tripod. Tripods were used as raised fire bowls, sympathetically drawing the sky's inspiration to earth.

115

Guardian Apollo

Figure 10.1

The white-and-black "harlequin" pattern highlights the forms of the Starcut. All the facets can be expressed in exact rational numbers and their fractions. All are bounded by lines that meet the sides of the square such that they divide the side into rational whole-number measures. All of them are prime-number divisions.

10
Conjecture

If a man's wit be wandering,
let him study the mathematics.

Francis Bacon

After investigating the seventh and eleventh cuts of the Starcut square, I suspected that it might be further divisible and asked a mathematician friend Elliott Manley if he could find any more odd and prime sections of the square's side that might arise from using slanted lines through the nodes of the lattice. The results were phenomenal.

The starting point for the construction (fig. 10.2, p. 118) is the division into elevenths.

This as we have seen, produces the points determining the Great Pyramid's profile. In the notebook page, reproduced as figure 10.3 on the next page, we can see how, continuing on from the eleventh marker points on the top edge of the square, lines ricochet back and forth. They always go exactly through the nodes of the diagram, they then cut the sides so as to give ever more exotic fractions (thirteenths, seventeenths, and nineteenths).

Elliot's simple procedure, using fractions and similar triangles, produced a profusion of ever-finer cuts of the square. He generated a spreadsheet of coordinates that demonstrated how to get all prime cuts from

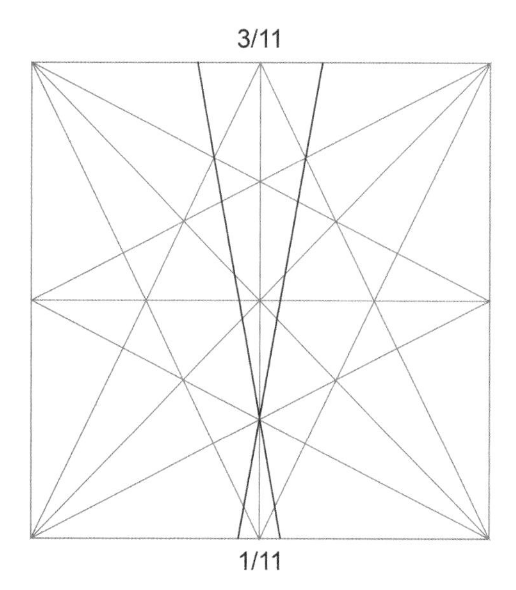

Figure 10.2
The starting point for the construction is the division into elevenths.

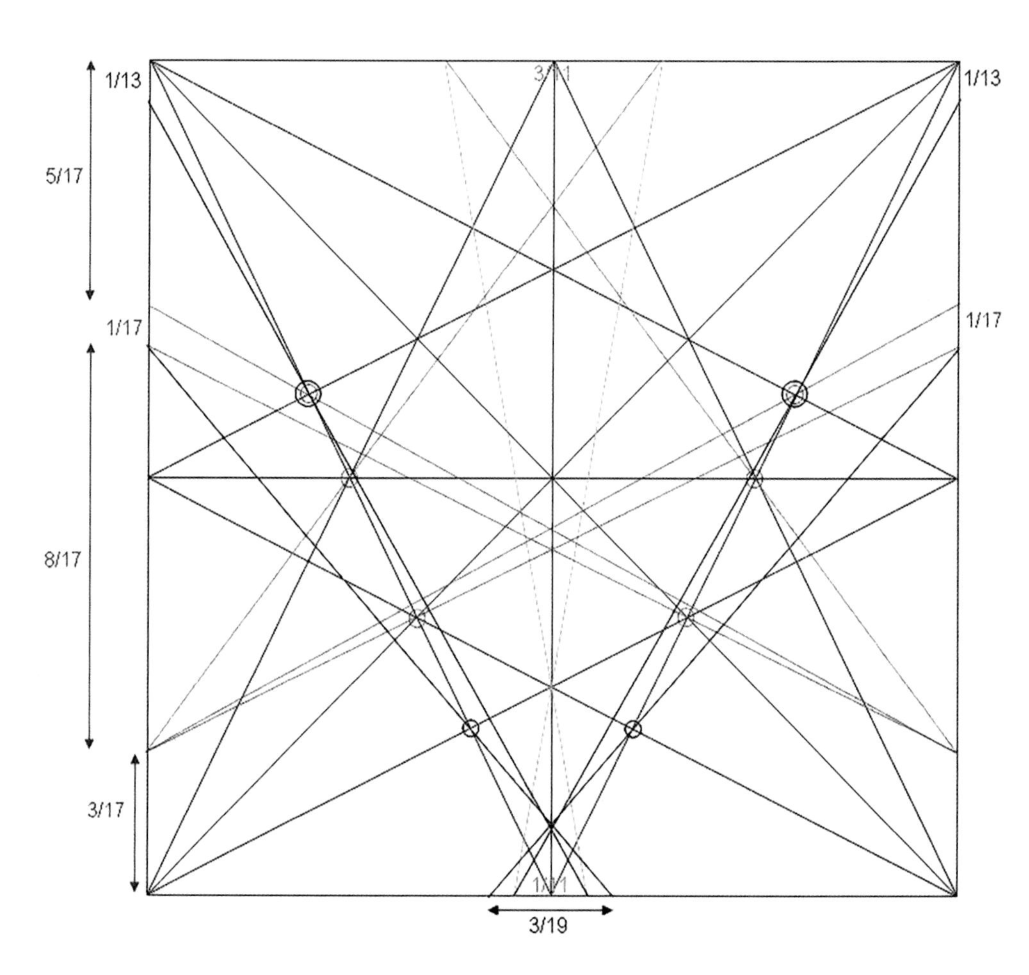

Figure 10.3
More exotic fractional cuts of the side result from additional lines that pass exactly through the nodes of the diagram.

thirteenths to "one-hundred-and-oneths"; hence our conjecture that

Conjecture

> Given a square with the centers of each side marked, all rational sections of the square's side can be found with straightedge only.

Of course this does not only hold for all *squares*. As was noted earlier in connection with using the Starcut form in designing stained glass, nodes of the lattice can be used in any *rectangle* so as to divide whatever length the sides happen to be into proportionally rational sections. And what is true for squares and rectangles also applies to parallelograms. The essential sectioning quality of the lattice remains unaltered whatever regular elongations or slanted lines are applied (see figs. 10.4 and 10.5). In these and the following figures the basic eleventh-cut lattice is used to illustrate the geometry.

The pattern has two further tricks up its sleeve. If the form in question is a trapezoid, the endless rational divisibility of the sides is retained—now with a perspective effect to the pattern (fig. 10.6).

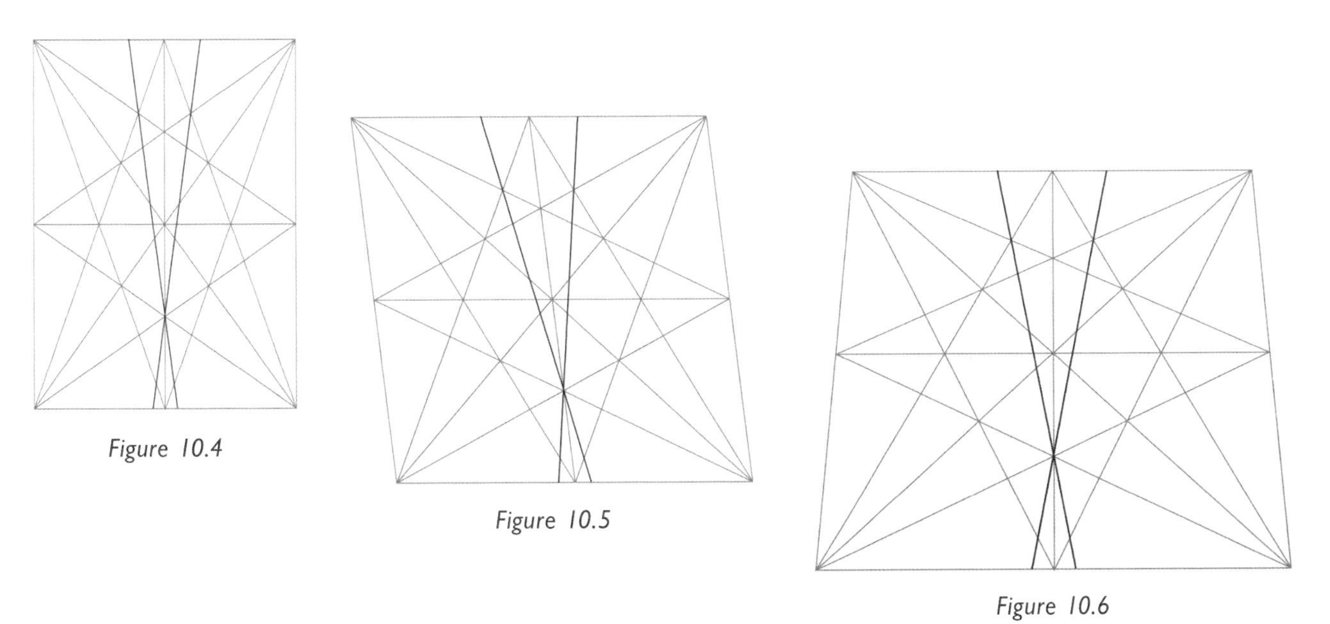

Figures 10.4–10.6. The essential sectioning quality of the lattice remains unaltered whatever regular elongations or pairs of slanted lines are applied, including when applied to a trapezoid.

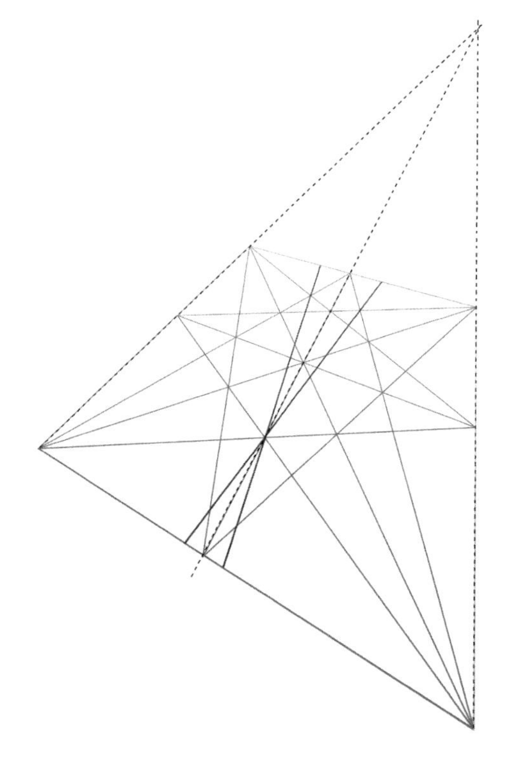

Figure 10.7
Illustrating that the starting point can simply be any quadrilateral of unequal sides of whatever lengths.

It emerges that, in fact, the starting point can simply be any quadrilateral of unequal sides of whatever lengths. The construction for placing the Starcut within the quadrilateral can be seen in figure 10.7.

First diagonals are drawn across from opposite corners of the quadrilateral to give a "center" point relative to the quadrilateral's outline. One then extends the two sides that are most inclined toward each other (they being the most convenient pair) until their extensions meet. From their meeting point a line is drawn through the quadrilateral such that it passes through the intersection of the diagonals. The points where this line cuts the sides of the quadrilateral are taken as guide points from which the Starcut can be drawn appropriate to the four-sided shape that one started with. One now has a figure that can be used—at least in principle—to subdivide infinitely and rationally the quadrilateral's sides, always according to an accurate perspectival projection.

The key throughout is that all lines, rebounding off the sides of whatever four-sided shape, must pass through one or more of the nodes of the inscribed Starcut.

It was way beyond my expectations that one could, at least in principle, get the guide points for *all* possible divisions of the sides of *all* such four-sided figures simply by using information integral to the figure's construction. The device had turned out to be a geometric genie.

In view of the fact that it can provide guidelines for all perspectival divisions of rectangular space, it is intriguing that the diagram should have first appeared in the historical record around the time that European artists were thoroughly investigating perspective.

There was also something else about the Starcut's historical significance—something that can be shown in a very simple way, as shown by the grid in figure 10.8.

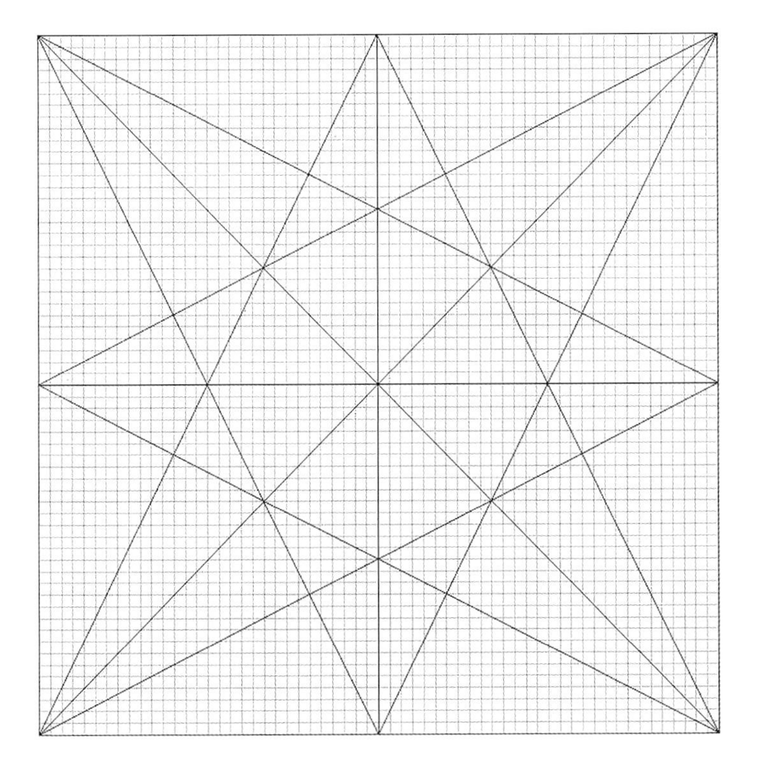

Figure 10.8
The ancient sexagesimal number system is pre-contained within the measures of the Starcut. This is demonstrated by the fact that all the nodes of the lattice only fall on the intersection points of a regular grid when that grid sections the square 60 × 60 (as shown here). This is because the nodes govern direct vertical and horizontal divisions by 3, 4, and 5—of which 60 is the product.

11
Beautiful Diagonals

What is the wisest thing? The wisest thing is number; the second wisest is the naming power.

Initial question and answer in the school of Pythagoras

The Starcut diagram is a genuine table of numbers—literally a surface that pre-contains its own numerical system in the pattern of its lines and the areas they determine. The interest starts with the two intersecting mid-side diagonals. They cut each other into segments of 1, 2, 3, and 4. This simple and seminal fact about mid-side diagonals I first came across in Robert Lawlor's brilliant book, *Sacred Geometry.*

These lengths tell us that the big sloping triangle is a 3-4-5 right-angled triangle. No wonder there is a theorem about squares and right-angled triangles; it turns out that the square itself is the natural habitat of the simplest of them. The Starcut figure, completed to include all mid-side diagonals, contains forty distinct triangles, thirty-two of which are right-angled.

The diagram itself will, as we shall see, suggest exact whole-number areas for all these triangles, but first we will consider their geometric forms.

Opposite: Figure 11.1 A modern statue of Pythagoras in the port of his home island of Samos.

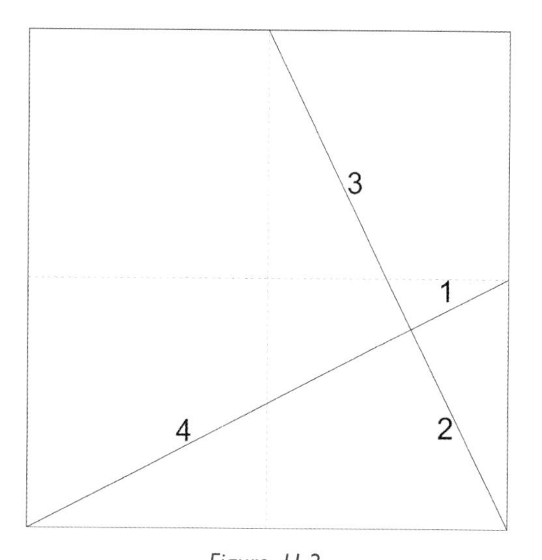

Figure 11.2
The two intersecting mid-side diagonals of a square cut each other into segments of 1, 2, 3, and 4.

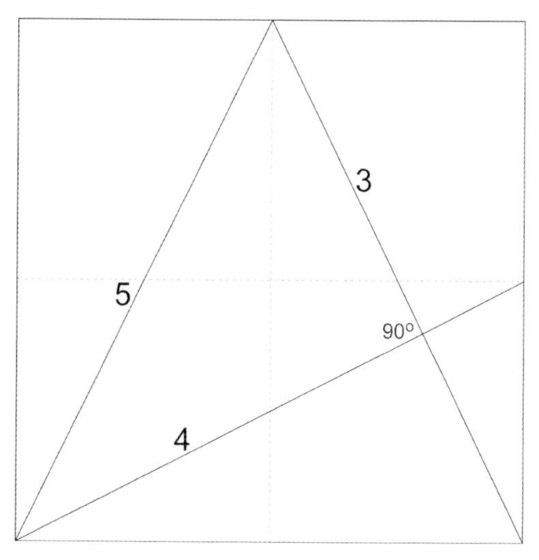

Figure 11.3
When a third mid-side diagonal is drawn the result is a 3-4-5 right triangle.

There are five sets of eight triangles in the whole pattern, as shown in figure 11.6. The innermost set, shown in figure 11.5, makes a semi-regular octagon composed of triangles with sides of different lengths ("scalene" therefore) and with different angles, all less than right angles. We will return to this set of triangles a little later.

The second set (fig. 11.7) makes an octagonal star of triangles with sides that are in 3:4:5 ratio to each other. They are scaled-down versions (1⁄36 of the area) of the big 3-4-5 triangle that was made when the first three mid-side diagonals were drawn (as was shown above in fig. 11.3). These are of course all right-angled triangles.

The third set (fig. 11.8) makes a four-pointed star of reflecting pairs joined at the meridians. The component triangles are of ratio 1:2:√5. That is a scaled-down version (1⁄20 of the area) of the triangle that appeared immediately when the first mid-side diagonal was made. That big original triangle is here outlined in bold. These too are all right-angled.

Note that the 1:2:√5 triangle gives an elegant version of the golden ratio within the basic lines of the Starcut diagram (fig. 11.9, p. 126).

Figure 11.4
With all the mid-side diagonals drawn there are forty total triangles in the diagram.

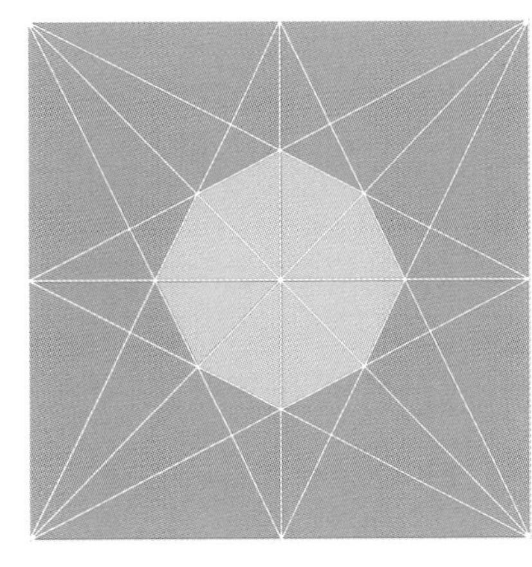

Figure 11.5
Only the central eight triangles are not right-angled.

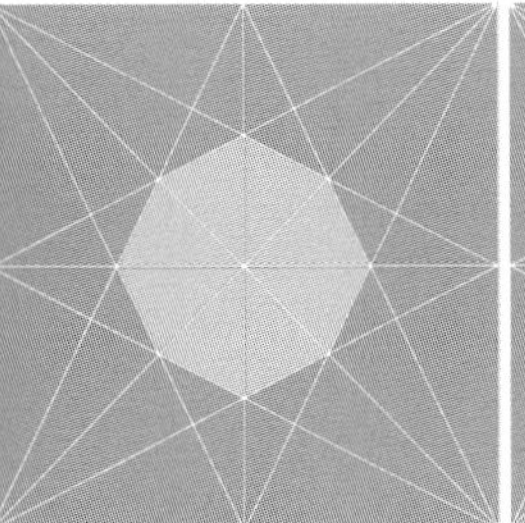

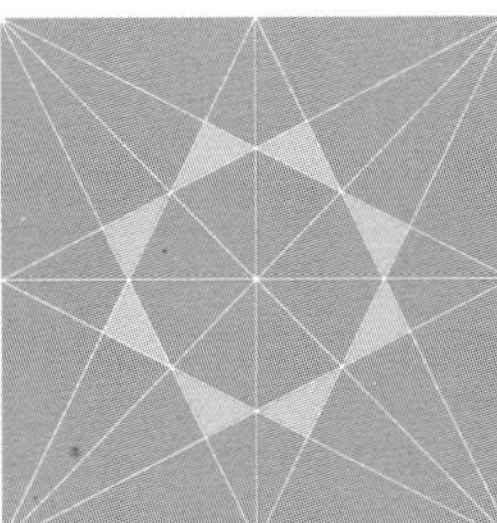

Figure 11.6
There are five sets of eight triangles in the whole pattern.

Figure 11.7
An octagonal star of triangles with sides that are in 3:4:5 ratio to each other.

Figure 11.8
The third set makes a four-pointed star of reflecting pairs joined at the meridians. Each of these triangles have sides in the ratio of 1:2:√5, which is the same as the larger bolded triangle that results from drawing a single mid-side diagonal.

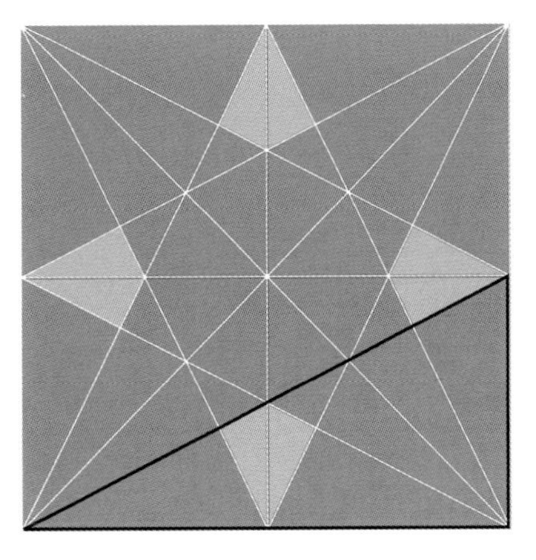

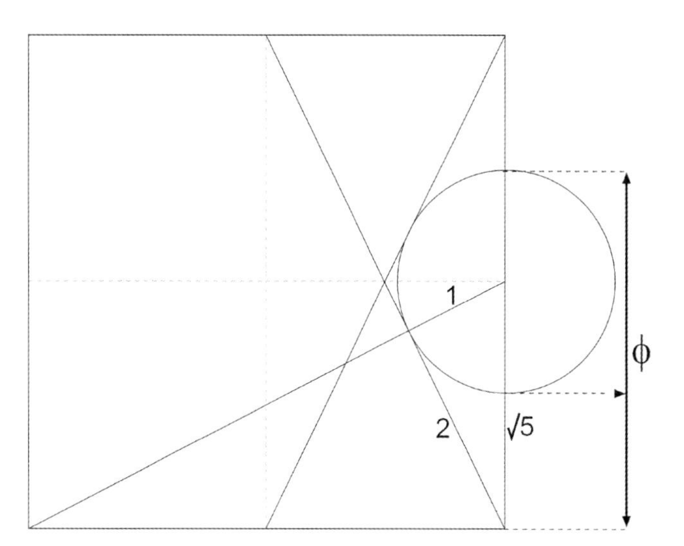

Figure 11.9
The 1:2:√5 ratio contains golden-proportion numbers: (√5 + 1)/2. A golden section ("cut") can be derived directly, using a circle of radius length 1 to cut the sides of the square—the arrow pointing to the side bar illustrates the ratio along the bar's length. The shorter section of the bar is to the longer part of it, as the longer section is to the whole bar.

The fourth set of component triangles in the Starcut is again made of reflected pairs of triangles forming a four-pointed star. This one is joined along the diagonals (fig. 11.10). These triangles are of side-length ratio 1:3:√10. They, too, are right-angled.

The fifth set, in figure 11.11, makes a boundary at the square's edge around the star form that is made by all the internal sets. These triangles are of the same shape as those in the third set, 1:2:√5 (see fig. 11.8), and are four times their area.

The last set to consider is that of the triangles that make up the central octagon (set 1 in fig. 11.6). These are not right-angled triangles. However, we can compare them with similar triangles, as is done here by highlighting one of the central triangles and darkening a similar triangle that is four times larger. They are then seen to be a combination of the two right triangles of the proportions 1:2:√5 and 1:3:√10.

SIMILAR TRIANGLES

When geometry first appears in the Mesopotamian record, the concept of similar triangles, where the same shape appears in different sizes, is already known. Certainly if the Starcut was, as I believe, one of geometry's earliest devices, the plethora of similar triangles that it contains would have made this seminal geometric principle obvious, and useful. In figures 11.13–11.16 (p. 128), we see how all the component triangles of the Starcut occur in various different sizes.

Figure 11.10
Reflected pairs of triangles forming a four-pointed star joined along the diagonals. The sides of each triangle are in the ratio of 1:3:√10.

Figure 11.11
The fifth set makes a boundary at the square's edge around the star form made by all the internal sets.

Figure 11.12
The highlighted central triangle is similar to the darkened triangle, which is four times larger.

Figure 11.13

Figure 11.14

Figure 11.15

Figure 11.16

Figures 11.13–11.16. In these four figures it is easy to see how all the component triangles of the Starcut occur in various different sizes.

There are many other qualities in the lattice. Later we will explore the way the smallest version of the 3-4-5 triangle provides a number system that rationalizes the entire figure numerically; for now, however, we may consider just one of its beautiful features.

The line that marks the threefold division of the square goes through the point of the smallest triangle that stands opposite its line of length 3 (fig. 11.17). An equivalent thing happens with the lines cutting the square by 4 (fig. 11.18) and by 5 (fig. 11.19). The symmetry means that in all eight small 3-4-5 triangles the same thing applies both vertically and horizontally.

Our starting point in all this has been the 3-4-5 triangle. This triangle, with its right-angled family of other triangles, is normally associated with mechanical and structural processes. From such came the first understandings of surveying and trigonometry and from that, in turn, came mechanics, engineering, an entire world of architectural and technological construction, geographic projection, ballistics, and even telemetry. Yet this number set that turns out to be the natural inhabitant of any square, and that we usually consider as only the special case that helps to prove the famous theorem about the square on the hypotenuse, is more than just a seed of mechanical culture. There is a

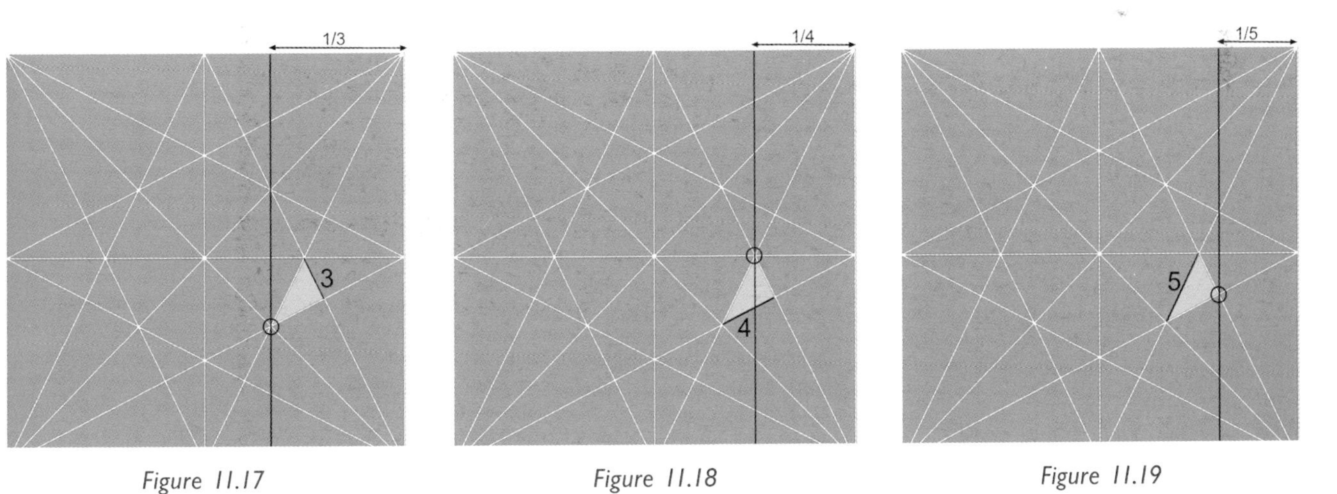

Figure 11.17 *Figure 11.18* *Figure 11.19*

Figures 11.17–11.19. The lines that mark the division of the square into thirds, fourths, and fifths pass through the points of the smallest triangle that lie opposite their respective line lengths.

quality of "life" about it that can be demonstrated, as we shall see in chapter 13, according to a powerful mathematical analogy. Just as there is a quality of "harmony" about it that will later be demonstrated in strict musical terms.

In the next chapter, however, having on more than one occasion suggested the Starcut as a possible unknown source of pre-Euclidean geometric lore, I feel it is necessary to show that, here and there in the ancient record and ancient practice, we find hints of its existence underlying a body of sophisticated knowledge whose origin we really know almost nothing about.

Figure 11.20

The two arcs of circles are drawn from the compass point positions ringed in black, such that they pass through the structural points ringed in green. The arcs cut the surrounding circle in such a way that the segment between the two red rings is $^1/_{56}$ of the circumference.

In his book *Solving Stonehenge: The New Key to an Ancient Enigma,* the archaeologist and surveyor Anthony Johnson proposes this construction as the probable method used to plot the positions of the fifty-six "Aubrey Holes" at Stonehenge.

It is believed that the Pythagoreans knew how to construct a polygon with fifty-six sides. It is a subtle method, closely approximate but not exact. It was proved by Gauss that there can be no exact geometric construction of any polygon whose sides are multiples of seven.

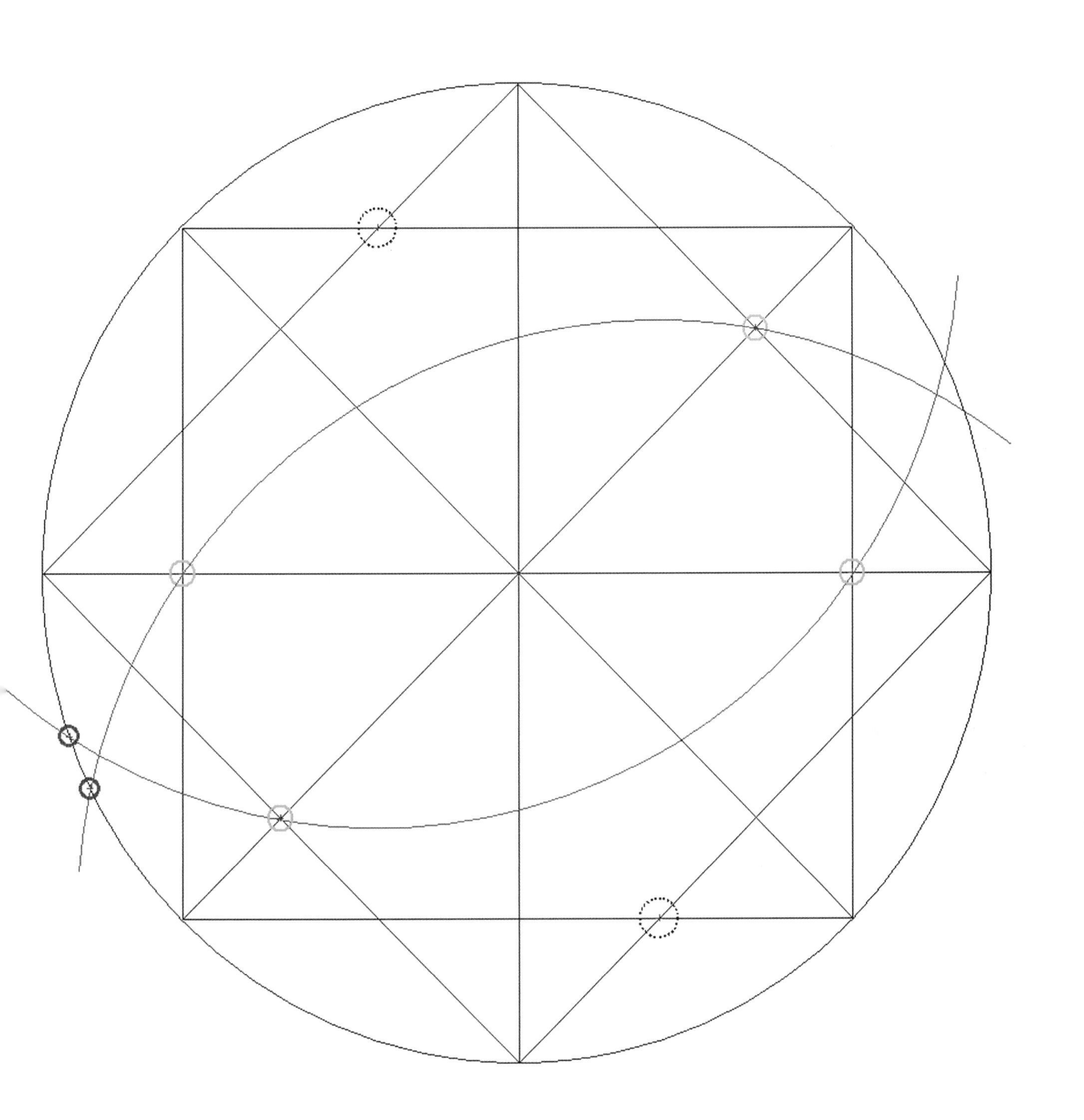

Figure 12.1
Stonehenge at sunset.

12
Clues to a Forgotten Lore

Plato and other initiated writers allude to a strange mystical science . . . no longer openly practised. . . . Its secrets were long preserved by the priests of old Egypt. It was, they said, the instrument by which they had maintained their civilization unchanged and with the same high cultural standards, for many thousands of years.

John Michell, introduction to *The Measure of Albion*, jointly authored with Robin Heath

The late John Michell and his associates have added hugely to the study of comparative metrology—the measurement systems of the ancient world. Any detailed account would take us too far from this book's theme; suffice it to say that they have marshaled evidence of remarkably accurate dimensions for the earth's polar and equatorial circumferences and the earth's mean radius. They have also published "Rosetta Stone" type listings by which the various measures of Egypt, Greece, Rome, Mesopotamia, and early Europe can be translated into each other. From all this work it becomes obvious that geometric understandings were very highly developed long before Euclid.

The great driver of early geometric study is indicated in the word itself: *Ge* (or Gaia)-*metria* = "Earth measurement." When our forebears began to settle in urban communities, and to cultivate the land to feed the numbers involved, there was an inevitable need for boundaries and measurement. It made practical sense to know the equivalent areas of differently shaped plots of land.

This must have been particularly the case in both Egypt and Mesopotamia where seasonal flooding was the rule rather than the exception. The ziggurats of Chaldea were not only defensive as possible fortresses against hostile attacks; they also offered refuge from the melt waters from the distant mountains and the overflow of the great rivers.

The importance of measuring celestial cycles was equally great—calendars are crucial to human societies. One of the best known uses of ziggurats was as platforms for observations of the sky.

However, beyond the strictly practical, there is clear evidence of a more exuberant intellectual/artistic interest in geometry and number. This evidence, in the form of intelligently conceived and executed structures, ground plans, and artifacts, is to be found right across the megalithic world that stretched from Orkney in the north, southward down the African coast to Cameroon, and eastward to western Asia.

Figure 12.2
This clay tablet dates to the fourteenth century BCE. It is a map of the countryside around the city of Nippur that was situated amid the branches of the Euphrates on the southern Mesopotamian floodplain.

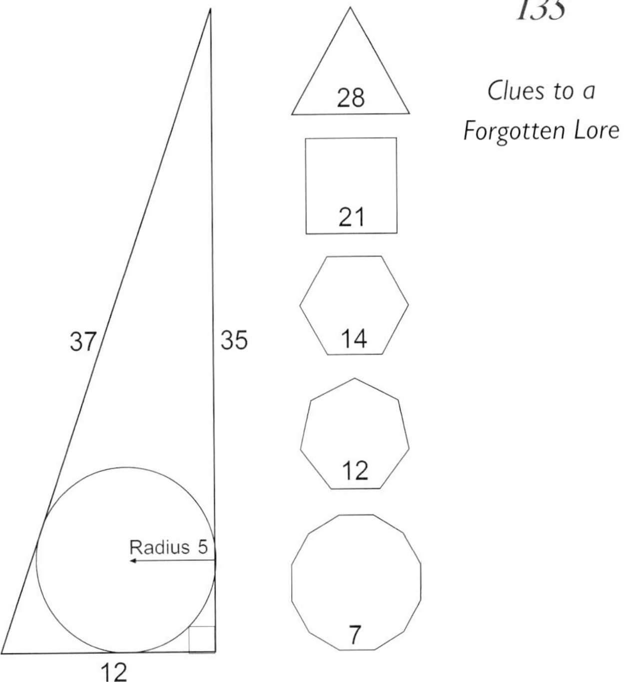

Figure 12.3
The triangle has an in-circle diameter of 10. Although the ancient world certainly did not have accurate degrees of angle measurements, I can't resist mentioning that this particular triangle has angles that are within hundredths of a degree of whole numbers: the larger being approximately 71 and the smaller approximately 19. The polygons, not drawn to strict scale, are included because the perimeter of this particular triangle, being 84, is rich in polygonal possibilities. It so happens that there is also a 3:4:5-ratio triangle with a perimeter of 84. Its sides are seven times the basic proportion, being 21:28:35.

WHOLE-NUMBER TRIANGLES

According to Professor Thom who surveyed at least five hundred British megalithic sites, their ovoid shapes seemed to be determined by swinging partial arcs from the points of whole-number right-angled triangles. His researches showed that the second most common such triangle (after the 3-4-5) was that with side lengths of 12, 35, and 37. It's worth a closer look at this triangle (fig. 12.3).

When giving presentations about the megalithic culture, I am often asked why whole-number proportions are so common. My assumption is that our ancestors were simply fascinated by the discovery of how such numbers worked together. It also seems probable that, just as the Pythagoreans were said to revere whole numbers as divine principles, others in the ancient world thought it natural to remain within their constraints.

Another strangely suggestive piece of whole-number triangle lore arises concerning what is sometimes called "Caesar's triangle." In the thirteenth chapter of book 1 of *De Bello Gallico,* Julius Caesar writes the following about Britain:

> The island is triangular in its form, and one of its sides is opposite to Gaul. One angle of this side, which is in Kent, . . . looks to the east; the lower looks to the south. This side extends about 500 miles. Another side lies toward Spain and the west, on which part is Ireland, less, as is reckoned, than Britain, by one half: but the passage from it into Britain is of equal distance with that from Gaul. . . . The length of this side, as their account states, is 700 miles. The third side is toward the north . . . but an angle of that side looks principally toward Germany. This side is considered to be 800 miles in length. Thus the whole island is 2,000 miles in circumference.

From the words "as their account states" one assumes that Caesar is quoting a British geographical record of some kind, perhaps even a written one. However there is something else that could be at work here, as is shown in figure 12.4.

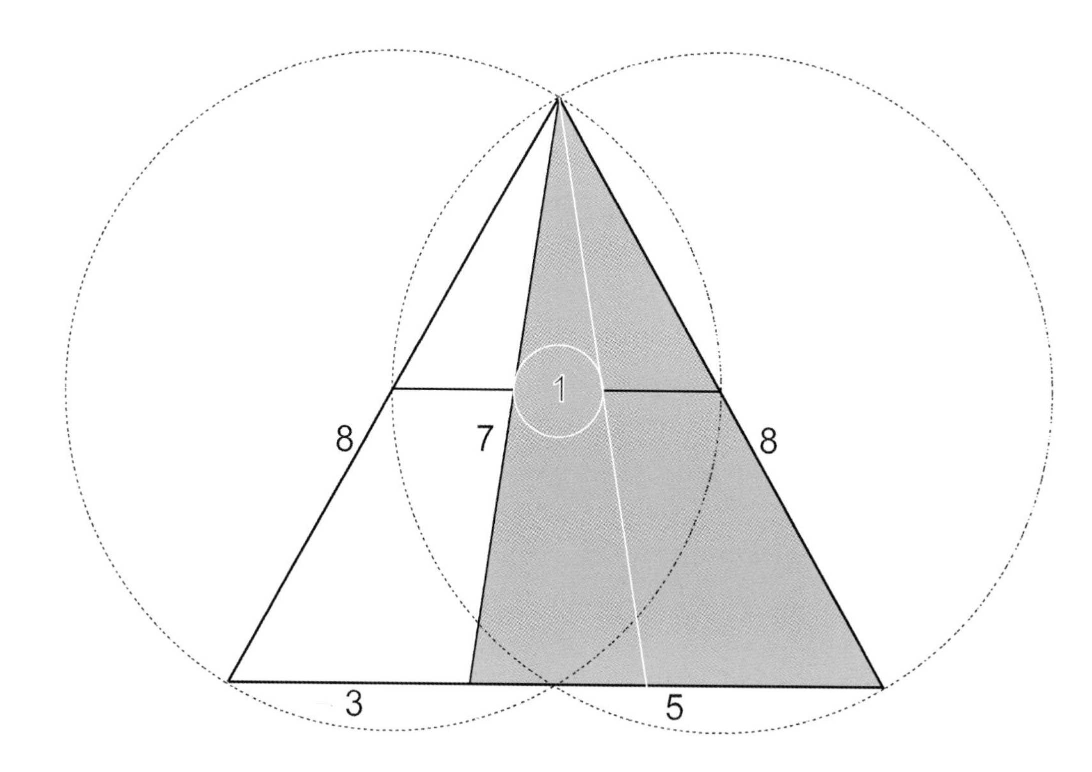

The ascribing of a triangle of such elegant construction (with its measures multiplied by one hundred) to the British Isles, would hardly have happened by sheer chance. It can only mean that a triangle of such proportions was an already known form as part of a developed tradition of whole-number geometry.

WHAT FIBONACCI DIDN'T KNOW

I came across indications of previous ancient knowledge of equilateral triangles in the work of the late Ben Iverson, an American mathematician who happened to mention this 5-7-8 structure in connection with tetrahedral and octahedral trusses used in modern bridge building. He claimed that knowledge of it formed part of a whole mathematical system that was used by, but predated, Pythagoras. Iverson called this system *quantum arithmetic* because it always dealt in whole numbers—*quanta*.

Opposite: Figure 12.4
The "triangle of Britain" turns out to be a form that arises naturally within the geometry of an equilateral triangle of side length 8. Within such a triangle a line of length exactly 7 can be dropped from the vertex to the opposite side, and it will be found to divide that side into exact whole-number lengths: 3 and 5. The extra white line in the diagram (also of length 7) and the small circle of diameter 1 are included to show the neatness of the construction that defines its own unit in this way. The radii of the governing circles are of length 4.

Iverson's work is a study in itself and warrants a fat book. Here I just want to refer to his assertion that this numerical proportion, within an equilateral triangle of side length 8, was considered the primary measure of all such triangles. Any equilateral triangle with a whole-number side length can be divided into rational lengths by a dropped line that is itself of rational length. The particular length of line in any given case can be calculated according to a general set of rules, applied to a system of four-number sets that are the basis of *quantum arithmetic,* and which Iverson calls "bead numbers." From these bead numbers a whole raft of verifiable conclusions can be derived about Pythagorean triple groups, equilateral triangles, ellipses, and much else that goes beyond our study here.

One key point emerges, however: When applied to right-angled triangles, exactly the same bead-number set that underlies the 5-7-8 section of the equilateral, gives rise to the 3-4-5 triangle itself—and that four number set is 1, 1, 2, 3, which are the first positive integers of what is now called the "Fibonacci series." This is interesting in our context because that number series is known to occur in growth patterns. It is generated by adding the two previous numbers to get the follow-

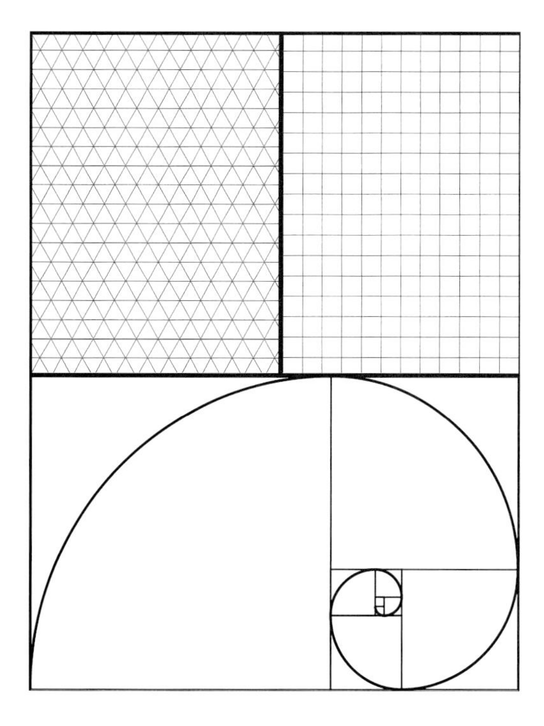

Figure 12.5
The normative all-space-filling grids of the triangle and square are here represented standing upon the Fibonacci spiral of growth created by drawing arcs connecting the opposite corners of squares in a tiling using squares of sizes 1, 1, 2, 3, 5, 8, 13, 21, and 34.

ing number and then repeating that process into a continuing addition sequence that quite literally grows out of itself. It is well known that as the Fibonacci series grows it generates a ratio between its last two numbers that approaches ever closer to the golden ratio—$1:(\sqrt{5} + 1)/2$.

It is remarkable that Iverson should have discovered such unifying arithmetic/geometric methods in the ancient world. Through them he has shown the linkage of the basic all-space-filling lattice of the equilateral triangle, with that of the square and its implicit 3-4-5 triangle and that of growing systems archetypally represented by the Fibonacci series.

So, taking into account the kind of geometric knowledge uncovered by Iverson; the kind of geometric knowledge that was familiar with an unexpected fifty-six-sided polygon; knowledge that developed a sophisticated but unexplained metrology of our planet; that explored the deployment of whole-number constructions on the land; that had applied such concepts both in geometric theory and in geographical accounts—one is indeed left with the sense that there is a huge gap in our overall appreciation of just what our ancestors did know, and how they came by their knowledge.

It is within that unknown cultural process that I believe the Starcut, with its simplicity of form, and rich educational potential, very probably played a seminal role.

> *The Mundus Imaginalis ["imaginal" world] . . . is the place in our mind where we store significant images. These form a hierarchy. . . . At the bottom are images that are personal and exclusive to ourselves. . . . Next come images that are shared within our culture. . . . Above this is the mythological level, where reside the images belonging to our particular religion. Next come the images shared by every human being such as parts of the body, mother, tree, mountain, and the time cycles that we visualize as day and year. Lastly are images that border on the abstract, like the tones, numbers, and forms studied by the Pythagoreans.*
>
> Joscelyn Godwin,
> *The Pagan Dream of the Renaissance*

Figure 13.1
Fromia monilis starfish.

13
The Mother of All Pentagons

From the right-angle scalene proceeds the genesis of all mundane bodies and the whole universe.

Athanasius Kircher, *Oedipus Aegyptiacus*

We will return to the Starcut's component forms and there find an unexpected relevance to cubes, and some mystical Kabbalistic lore that goes with them. Here, however, I want to revisit the topic of number as "quality," and expand on the qualities that are present in the diagram. We start with the bare-bones "four-ness" of the Starcut.

A square is a fourfold figure: 4 sides, 4 points, and $2 \times 4 = 8$ peripheral points, 4 of them at the meridians and 4 at the corners joined by diagonals, which themselves appear in 4 sections. The square is also cut into 4 by a central right angle—and so on. The common subdivision shown in figure 13.3 on the next page essentially continues that four-ness by a process of doubling.

What the Starcut subdivision does, however, is to bring five-ness into the game.

We have seen how the mid-side diagonals establish a point of intersection from which an entire number system can be derived without

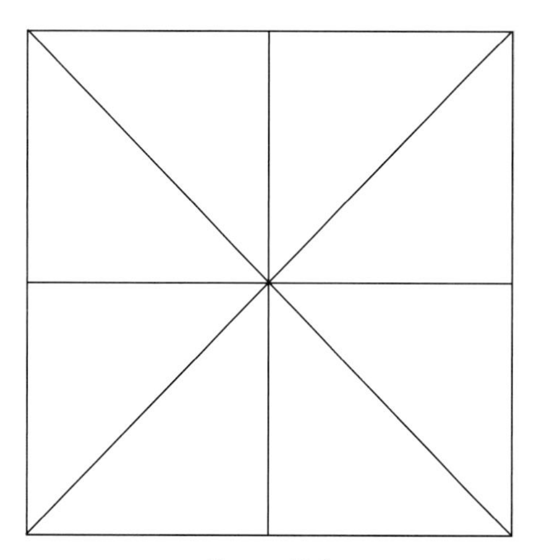

Figure 13.2
A square is a fourfold figure.

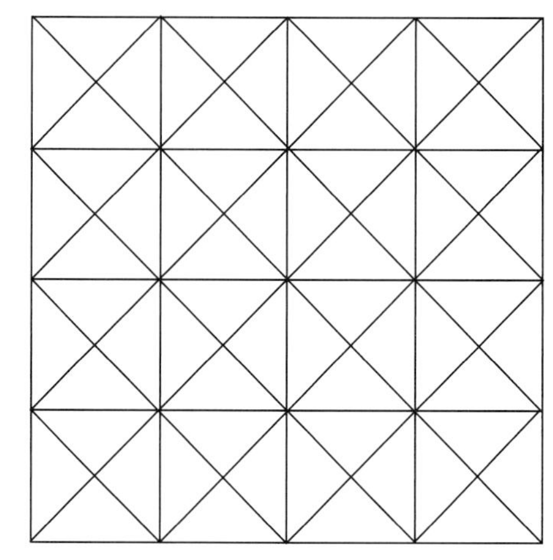

Figure 13.3
Four-ness is continued through doubling.

bringing anything in from the outside except the ability to measure sameness and some way of naming measured amounts, which is an excellent example of the Pythagoreans' "wisest and the second wisest things"—number and naming.

In figure 13.4 the right-angled intersection point of the two mid-side diagonals opens up a "five-ness" latent in the "two-four-eight-ness" of the square. The mid-side diagonals have a rationale of 5. They divide each other exactly into 1 + 4 and 3 + 2. Each has a length of 5, and fivefold divisions are spelled out in their relative measures. It takes two such lines to reveal this, giving the 10; it takes a third mid-side diagonal to draw the 5 as the hypotenuse of the 3-4-5 triangle.

The interaction of geometry and number is beautifully illustrated when one "womeks" the generation of the complete lattice from its beginning, as in figure 13.5. It originated in the circle, becoming two-fold in the vesica. Threefold-ness emerged from the resulting rhombus shape being made up of two equilateral triangles, one of which is shown (shaded) here. There was already an implied sixfold-ness in the common radius' division of the circumference of both vesica circles. The four had immediately appeared as the four defined points of the vesica rhombus producing the central right angles (those perpendicular bisectors!).

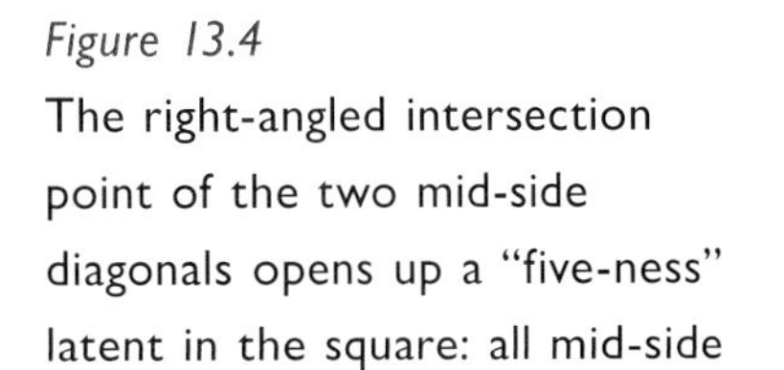

Figure 13.4
The right-angled intersection point of the two mid-side diagonals opens up a "five-ness" latent in the square: all mid-side diagonals have a length of 5.

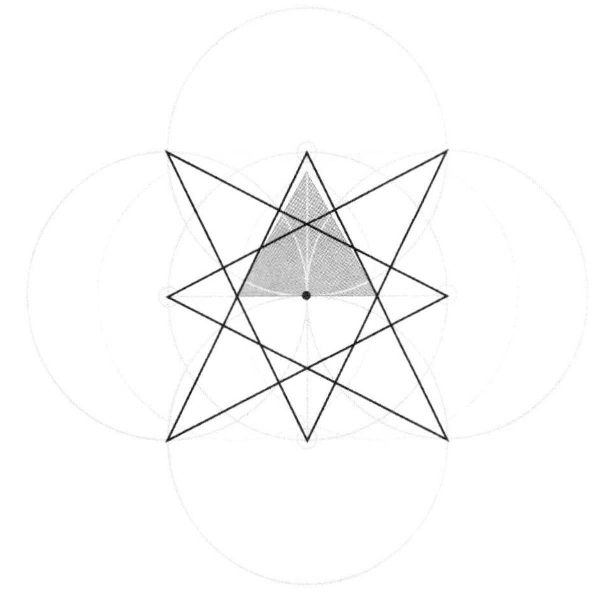

Figure 13.5
The Starcut retaining the geometric steps of its generation.

There were in all seven constructional circles giving the nine anchor points of an eightfold star.

And here—in the linear connections of that star—lies the five. The quality of five is integral to the immense productivity of the Starcut, because five—with its relations √5 and the golden number phi (ϕ)—has the quality of life. Many authors have written on the golden number/section/mean, on its connections with the pentagon, with logarithmic spirals, organic growth patterns, and the association between five-ish geometry and life.

In Plato's *Timaeus* he speaks of the very first stirring, the very first differentiation occurring in the uncreated, and thus divine, unity. It seems to me that the golden ratio would be the focus of the first tremor of such differentiation. An octave comes from twofold geometry, to do with doubling and halving (as in fig. 13.3 p. 142). We will be seeing and hearing more of octaves later. The interval of an octave upward along a sounding string is produced by the half-length of that string. However, the interval of a "golden section" of a string is not an exact note in our musical scale, nor is it an exact number. Its decimal starts as 0.6180339 . . . and goes on forever without the repetition of any pattern. So there is no exact solution to $(\sqrt{5} + 1)/2$, it is irrational. This is so because $\sqrt{5}$ itself is irrational.

As geometry, however, the golden ratio is spot on—it is distinguished by a point that divides the length without losing its inner coherence—because its shorter part is to its longer part as the longer is to the whole line.

The kind of incremental process indicated in the geometry in figure 13.6 incarnates in profusion throughout the living world. If one

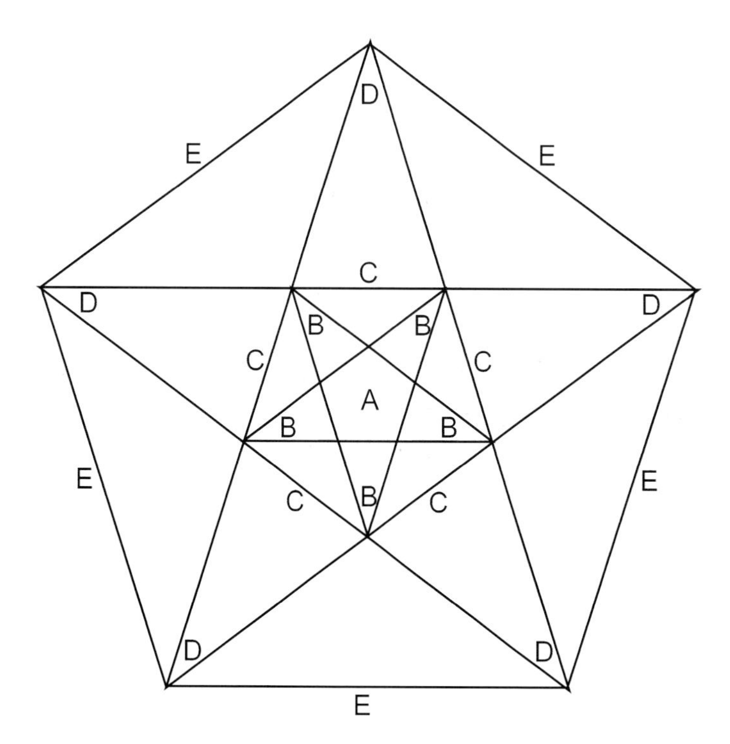

Figure 13.6

The innermost pentagon has sides of length *A*. The extensions to points of the pentangle star are of length *B*. *A* and *B* are in a golden-ratio relationship. So *A* is to *B* as *B* is to *A* plus *B*. *B* is in golden-ratio relationship to *C*, which is the side length of the pentagon that encloses the inner star. *C* is in the same ratio to *D*, the extensions to the points of the big star; and the ratio is again repeated relating *D* and *E*, the side length of the outer pentagon.

imagines all these lengths as a progressively unfolding series, one has a process model that is characteristic of growing life. Figure 13.7 reflects it perfectly.

In order to draw a pentagon at all, one needs to use a square's mid-side diagonal via one construction or another. In figure 13.8 (p. 147), I have boxed the circle in which the pentagon is to be drawn and included a mid-side diagonal. It is this line that is taken as a radius and initiates the pentagonal construction by being swung down to the meridian line to establish a point upon it, as in figure 13.9. We have met this

Figure 13.7
A columbine.

construction before in “Pyramid Connections” (chapter 6). This point along the meridian line produces a golden ratio in relation to the radius of the circle. A new radius is set, from the top end of the same mid-side diagonal to the point that has been made on the meridian. An arc is then swung to cut the big circle’s circumference into exact fifths to either side of the upper central point, as in figure 13.10. This length can be stepped around the circle giving the other pentagonal cuts shown in figure 13.11.

It is that crucial mid-side diagonal of the square that brings in the fivefold potential.

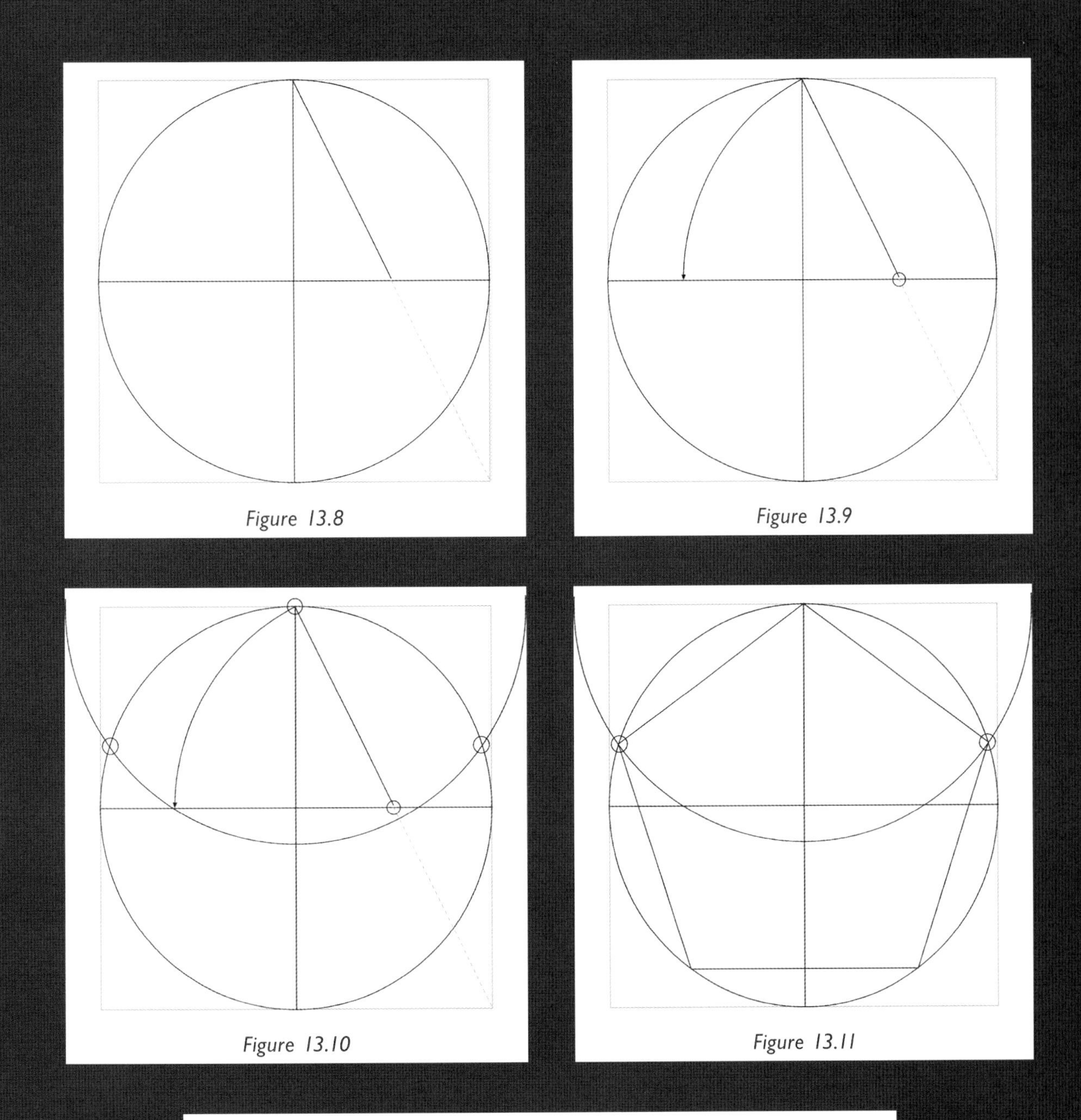

Figure 13.8

Figure 13.9

Figure 13.10

Figure 13.11

Figures 13.8–13.11. Steps in the construction of a pentagon starting from a mid-side diagonal.

Figure 14.1

The Triangle of Gods: According to the Egyptian priests, the "3 side" of the 3-4-5 triangle denoted Osiris (odd-numbered, male); the "4 side" denoted Isis (even-numbered, female); and the "5 side" was their offspring Horus, son of heaven, incarnate in the pharaoh. A twelve-knot cord (3 + 4 + 5, giving a right angle) is said to have been used to resurvey boundaries after the yearly Nile flood. Here, in the center of the montage, is included Seshat, the patron *neter* (a personified principle rather than a goddess) of measures and the cord-stretching rituals of Egypt. A female priest would represent her, advising the pharaoh, in the measuring ceremonies for temple ground plans. Here with her right hand she is using a counting stick that has seventy-two divisions (not all visible). Her left hand is holding the measures at the halfway (thirty-sixth) point down the calibration.

14
The Area Key

The first number system to emerge was when the two intersecting mid-side diagonals cut each other into sections of 1, 4 and 2, 3—that is, two lengths of 5 (fig. 14.2, p. 150). It took two lines and therefore a total measure of 10 to define that information. With a third mid-side diagonal, a big triangle in 3:4:5 ratio was revealed. The area of that big triangle was 6—a triangle's area being half its height times its base.

A second number system emerges quite naturally within the diagram since the smallest triangle in the Starcut is also of ratio 3:4:5, and we can conveniently take these three numbers as actual lengths to give us a more detailed measuring system for the entire lattice. This immediately gives us an area of 6 (shaded in fig. 14.3) for that smallest triangle. The other areas and lengths in the diagram are then very easily calculated. The entire figure decodes itself numerically. The second layer of numbering multiplies the line lengths by 6, and the areas by 36—that is, 6^2.

The one-eighth slice (or "octant") of the whole square marked in figure 14.3, labeled 15, 6, 9, 24, and 36 is the key to all areas. Each octant therefore sums to an area of 90. Thus the whole square, "on its own terms," has an area of 720. The length of the whole mid-side diagonal is now seen to be 30—Cesariano's measure (see chapter 7). As can be seen in figure 14.4, the side length of the main square and its ordinary diagonals, as opposed to the mid-side diagonals, are now all square roots of simple whole numbers.

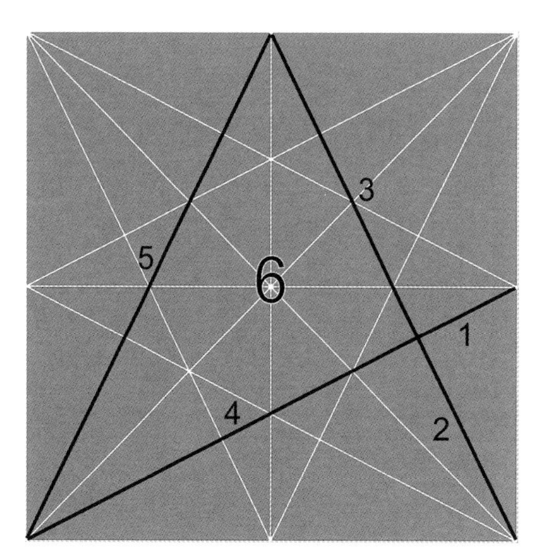

Figure 14.2

Two intersecting mid-side diagonals cut each other into sections—1, 4 and 2, 3—that is, two lengths of 5. A third diagonal gives a 3-4-5 triangle whose area is 6.

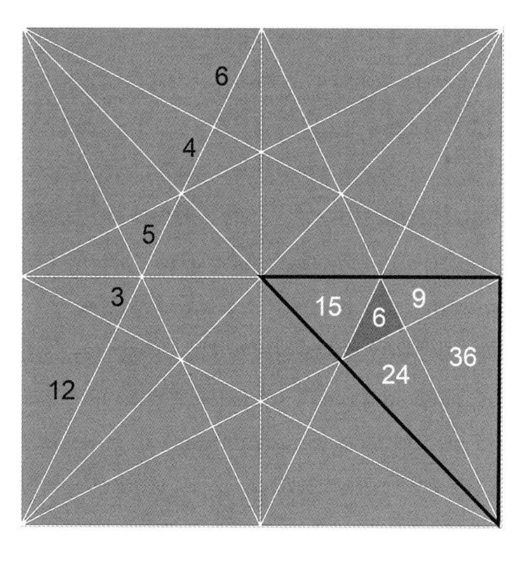

Figure 14.3

A second number system emerges from the smallest 3-4-5 triangle. The black numbers are the lengths of sections of the mid-side diagonal; the white numbers are areas. The areas of the triangles in the one-eighth slice of the whole square are the key to all the areas in the diagram.

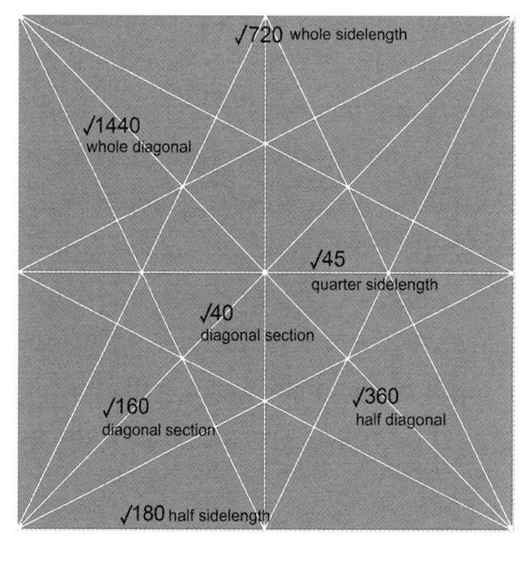

Figure 14.4

The side length of the main square and its ordinary diagonals (and sections of them), as opposed to the mid-side diagonals, are now all square roots of simple whole numbers.

Figure 14.5
Overlapping 3-4-5 triangles combine to make the complete Starcut form.

The entire lattice can now be seen to be a nest of overlapping 3-4-5 triangles that combine to make the complete Starcut form (fig. 14.5). They are separated out and shown in figure 14.6. There are four sizes of such triangles. They exist in sets of eight; there are thirty-two of them in all.

Pythagoras's name is associated with the right-angled triangle theorem, though it had been proved in Mesopotamia, India, and probably China before his time. His name is more exclusively associated with "perfect numbers" where a number is the same as the total of its factors. The number 28, the total of 1, 2, 4, 7, and 14, is an example.

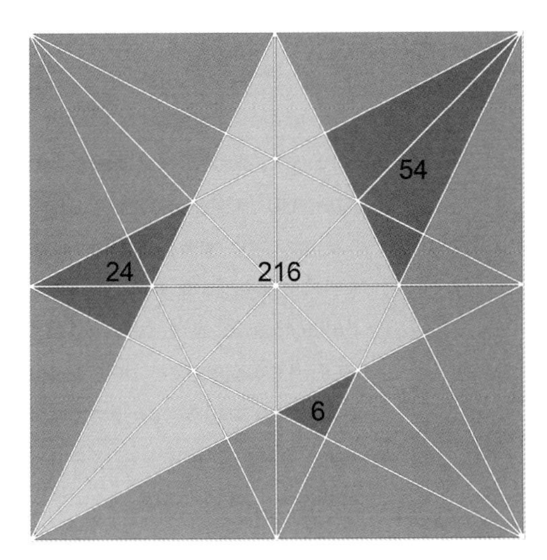

Figure 14.6
The four sizes of 3-4-5 triangles in the Starcut. Notice that all the angles touching corners and mid-sides are acute angles from the 3-4-5 triangle. When the areas of these triangles are divided by 6, the squares of the first perfect-number series are obtained: 1^2, 2^2, 3^2, and 6^2.

The first such number is 6, which is not only the sum of its factors 1, 2, and 3, but is also the multiple of them ("omni-perfect" therefore?). Here in the Starcut square there is an exact coincidence of these two information sets: one about right-angled triangles and one about the first perfect-number series. This is because the areas 6, 24, 54, and 216, when divided by 6, give 1, 4, 9, and 36. These are the squares of the first perfect-number series: 1^2, 2^2, 3^2, and 6^2.

This perfect-number series built into the natural measures of the Starcut was one of my very first observations about the whole numerical aspect; and it was this—plus the fact that the Starcut makes the theorem so obvious—that first led me to suspect that Pythagoras was aware of the diagram.

The perfect-number series is also there in a set of cubes.

THE CUBE

The primary elements of the geometry immediately imply the most rudimentary example of any triangle—a triangle of unit area, as is shown (shaded) in the bottom right corner of figure 14.7. The first two intersecting mid-side diagonals result in that triangle whose sides are in the ratio 1:2:√5. Thus its area, in terms of our first number system, is 1. The other obvious area in this "macro" version is that of the big triangle, which is 6. That is the surface area of a cube that has a volume

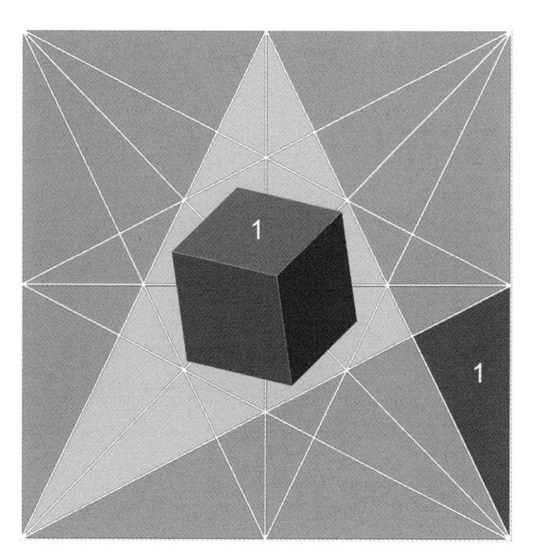

Figure 14.7
A triangle of unit area with sides in ratio of 1:2:√5 is shown (shaded) in the bottom right corner. The area of the larger 3-4-5 triangle is then 6, the surface area of a cube with a face of 1. This suggests unit spaces in both two and three dimensions: along with a unit triangle is implied a unit cube.

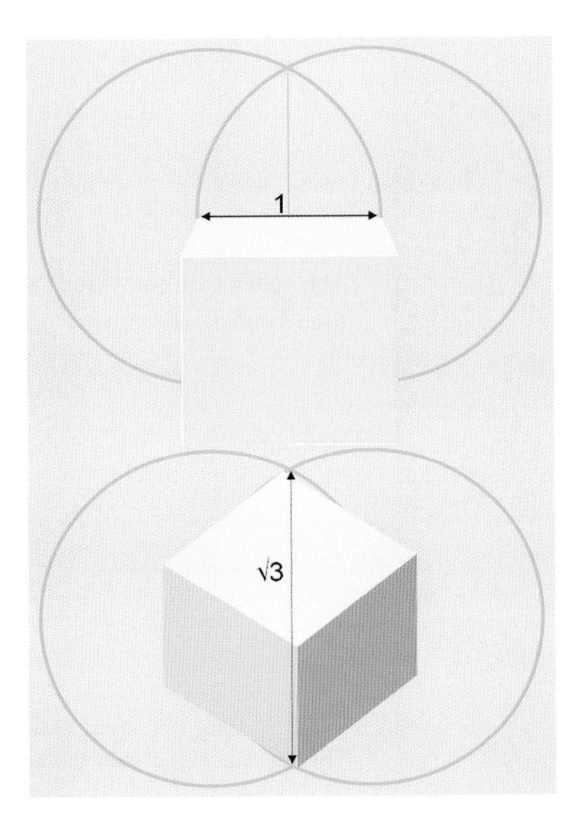

Figure 14.8
If the circles' radius is taken as 1, then the length of the vertical line between the points where the circles cut each other is √3. The ratio 1:√3 is exactly that which exists between the side of a unit cube and the diagonal from one corner to its opposite, passing through the center of the volume.

of 1 unit. The diagram therefore suggests unit spaces in both two and three dimensions: along with its unit triangle is implied a unit cube.

The vesica piscis also has this quality of suggesting measures in both two and three dimensions. This is shown in glorious Technicolor in figure 14.8.

Returning to the more detailed numbering system that comes out of the complete lattice where the smallest triangle has an area of 6—this can then be seen (in fig. 14.9) as the surface area of a scaled-down unit cube. The next bigger triangle, of area 24, has the surface area of the cube of 2, each face being of area 4. The area 54, of the next bigger triangle, is the surface area of the cube of 3, with each face being 9, and 216 is the surface area of the cube of 6—each face being 36.

In addition to the 216 area of the big triangle, the eight-pointed star has a value of 432 (shown in fig. 14.10), which, when added to the eight surrounding triangles, each of which has an area of 36, gives the entire square an area of 720.

Figure 14.9
The cubes that correspond to the four 3-4-5 triangles.

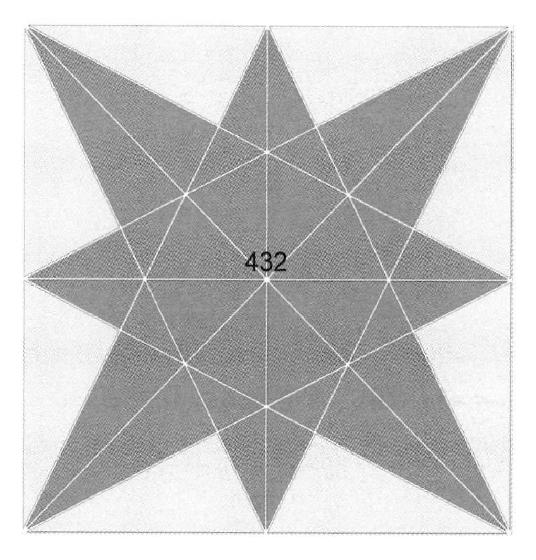

Figure 14.10
The eight-pointed star has an area of 432. When added to the eight surrounding triangles, each with an area of 36, the entire square is shown to have an area of 720.

These three values 216, 432, and 720 (to say nothing of others that are to be found in the lattice) immediately pitch us into direct associations with ancient canonical measures and the whole topic of esoteric number, which is developed in connection with the Kabbalah and Eastern traditions in subsequent chapters.

It becomes increasingly difficult not to see in the Starcut diagram a plausible matrix of ancient thought, both mathematical and metaphysical. Either that, or this simple diagram really is one of the greatest flukes on earth.

Figure 14.11
This woodcut is from Bernard Granollach's *Lunarium* of 1491; the text on the scroll reads: "The Higher Spirit resides [literally 'in-cubates'] beneath the appearance of the earthly."

Let the testimony of geometry be the Tabernacle that was constructed, and the Ark that was fashioned, formed in most regular proportions, and through divine ideas, by the gift of understanding, which leads us from things of sense to intellectual objects, or rather from these to holy things, and to the Holy of Holies.

FROM THE *STROMATA* OF
CLEMENT OF ALEXANDRIA

Figure 15.1
Ruins of the ancient synagogue in Kfar Bar'am, which dates from the third century CE.

15
The Secret and the Sacred

God is the circle, the square and the triangle, the center and the line—all things to all.

PHEREKYDES OF SYROS,
TEACHER OF PYTHAGORAS

We have found that the area of the central triangle of the Starcut is 216 (see fig. 14.6, p. 151). Perhaps Vitruvius only got part of the Pythagorean story that 216 was the perfect number of the cube. It really comes into its own when it is considered as an area value rather than a number of lines, A few years ago, I saw a mathematics textbook in which 216's perfection, as an example of Pythagoras's "number mysticism" was mocked as meaningless. The writer of that textbook had, ironically, let his anti-mystical prejudices interfere with some sums of which even Mr. Gradgrind would have approved. Such as the following:

- 216 is $6 \times 6 \times 6 = 6^3$.
- It is the sum of 3^3, 4^3, and 5^3.
- A cube of volume 216 (6^3) also has a surface area of 216; this cube is unique in having such an area-to-volume equivalence.
- Also, since there are three right angles at all eight corner points of a cube there are $24 \times 90°$ in the volume, amounting to 2,160°.

- In planar terms, a cube is the unique arrangement of six similar planes each having four equal sides, establishing a volume with nine planes of symmetry; 216, as the product of 4 × 6 × 9, is the perfect mnemonic for that information also.

So 216 *is* quite emphatically the number of the cube in plain Gradgrind terms.

However, here things do get considerably deeper. Arabic, Greek, and Hebrew all used letters for numbers, with the effect that they could be used for encoding additional concepts around a single word, because equivalent combinations of letters or numbers could be derived by various forms of substitution and permutation. This has given those languages a sacred terminology that is prismatic or multifaceted. This may exasperate the rationalist, but it provides artists and mystics (and sacred geometers) with a powerful and flexible symbolic language.

In the Jewish Kabbalah, the word *D'BIR,* meaning "oracle," has a value of 216: דביר (D'BIR reading right to left).

In the basic number-substitution code, or *gematria,* daleth (*d*) has a value of 4; beth (*b*) is 2; yod (*i*) is 10; and resh (*r*), has the value of 200. The whole word's value was thus 216.

The D'BIR, the "oracle" itself, was the Holy of Holies, the sanctuary of the "Presence." This was the most sacred place in all Judaism, the central cubic chamber of the Temple. Only the high priest was allowed in that cube—and even he, only once a year. Only there could the holiest Name be spoken and that too only once a year.

Figure 15.2 is a diagram based on the measures of the Temple given in the first book of Kings. Notice the raised cube of the Holy of Holies, in the upper left side, which was presumably approached by the high priest up a stairway.

> Solomon prepared the inner sanctuary within the temple to set the Ark of the Covenant of the Lord there. The inner sanctuary was twenty cubits long, twenty cubits wide and twenty high. (1 Kings 6:19)

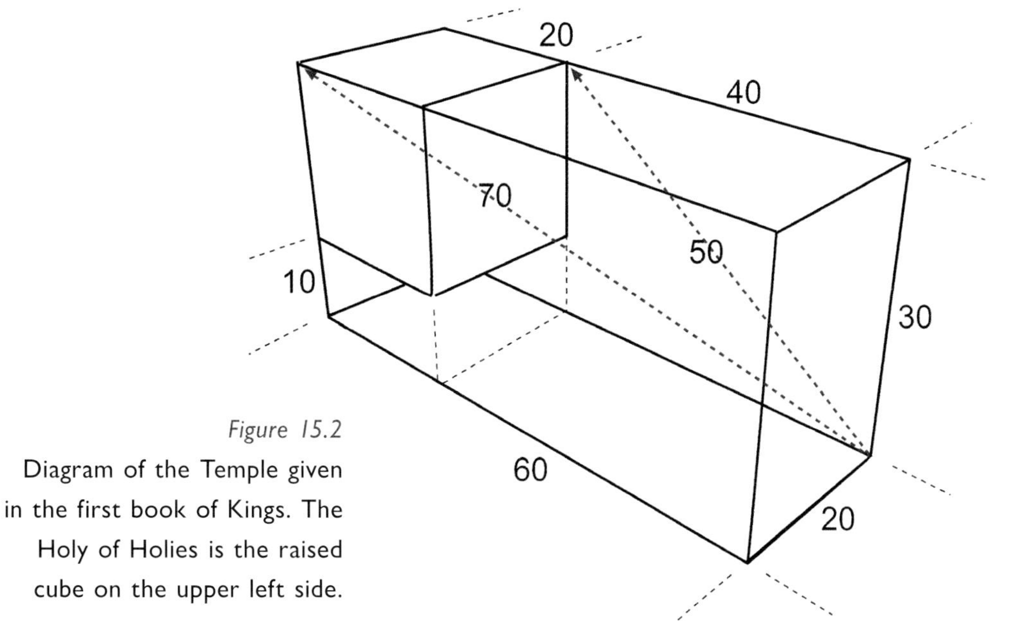

Figure 15.2
Diagram of the Temple given in the first book of Kings. The Holy of Holies is the raised cube on the upper left side.

It is immediately obvious that the numbers of the word *D'BIR* are not the numbers given in the written account, where the holy place is indeed a cube but measures 20 cubits in each direction. We will come back to that shortly. Notice in figure 15.2, which indicates the rest of the main Temple structure, that the distance going from one lower corner diagonally across the space to the opposite upper corner is 70. This measure is not mentioned at all in the passage in the first book of Kings. Since 7 and 70 are always sacred measures in Judaism and most other traditions, it is notable that it is not spelled out in the Temple description and is only discovered if one calculates on the basis of the measures that are given. All the measures 1 to 7 are incorporated (as multiples of 10). It is a beautiful structure of whole-number geometry.

The inner sanctuary then is a cube of volume 8,000 cubits (20 × 20 × 20) and with a surface area of 2,400 (20 × 20 × 6). Here, remembering that the first book of Kings may well have been compiled during the lifetime of Pythagoras (sixth century BCE), I have taken the liberty of performing the operation that Pythagoras himself performed in comparing the volume and the surface area of his cube of 6. The ratio here, between 2,400 and 8,000, is 6:20. This can be further reduced to 3:10 but there is a reason for leaving it as 6:20 that becomes apparent when we return to our first elemental division of the square by its mid-side diagonals.

The 216 area comes from the numeration suggested by the completed lattice, but this array reminds us that, in the first instance, the diagram gave us a "macro" number system. The areas from that system are shown in figure 15.3. The big triangle, having 3-4-5 side lengths, has an area of 6, and the other areas are simply calculated, showing that the whole square has an area of 20. The ratio, 6:20, now reflects that between the volume and area of the oracle according to the measurements given in the Bible. This seems simply a coincidence.

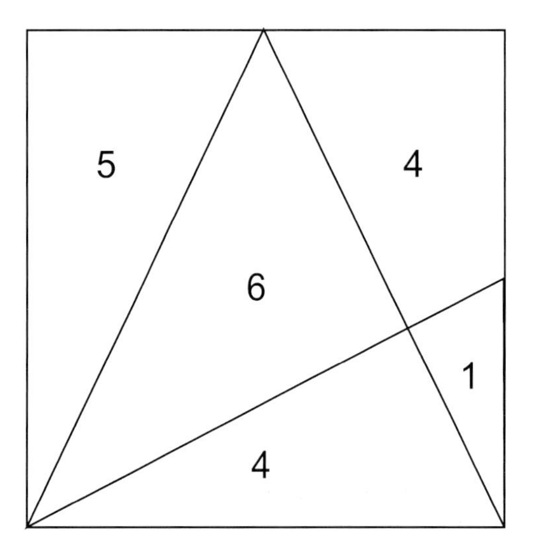

Figure 15.3
"Macro" number-system areas calculated from the first 3-4-5 triangle with area 6, yielding a square with an area of 20.

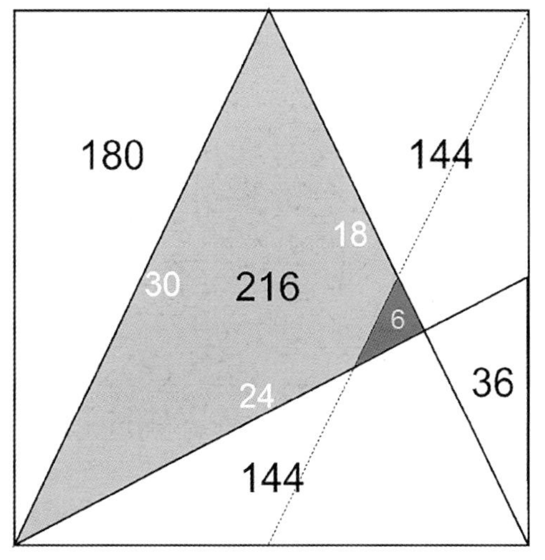

Figure 15.4
The second number system populated from the smallest 3-4-5 triangle emphasizes the number 144, a number with connections to Jewish and Christian traditions.

If we turn, however, to the second number system that we took from the smallest 3-4-5 triangle, the system that rationalized the whole square into an area of 720, we find a nest of coincidences (fig. 15.4). The most noticeable is the emphasis of 144. This multiplied by 1,000, is, in both Jewish and Christian traditions, the number that symbolizes the tribes of the blessed—12,000 from all twelve tribes. Tradition holds that this was the number of the Israelites when they first entered the land, crossing the Jordan at Gilgal. The same number occurs in Revelation as the number of those washed in the blood of the lamb. However, the more intriguing coincidence, I believe, is to do with the central triangle of area 216—the number of the sanctuary.

With the 216 (D'BIR) we are dealing with the Kabbalah rather than with the public Torah. Its triangular 3-4-5 version has a perimeter of length 72 (30 + 24 + 18), which is also the edge length of its cubic version since the cube sides are of length 6 and there are 12 of them. The number 72—especially in association with 216—takes us to a core Kabbalistic tradition: that of the seventy-two names of God.

Each name contains three letters. So the seventy-two contain 216 letters (72 × 3, fig. 15.5). The elemental triangle of the Starcut is bounded by the number of Divine Names, which further confirms

Figure 15.5

The seventy-two Divine names of God.

the number 216 that is suspended in a space—the area of the whole square—that is itself a measure of 72 in decuple form (720). The symbol systems mesh like a palindrome.

Reference to the Names brings us to the fact that the living—as opposed to written—Kabbalah, is an oral tradition. Apart from any reasons of esoteric confidentiality, the tradition has to be oral because the use of the Divine Names is integral to its practice, and both the pronunciation and the intonation are passed on "mouth to ear." The Hebrew language contains its own guttural, lingual, labial, sibilant, and palatal sounds as well as the aspirate "holy vowel breaths" that were not written. Further yet, the sacred words were sung. It is significant that one Kabbalistic tradition states that if the pupil can recall the voice of the teacher, a perfect spiritual telepathy can be achieved. Not for nothing does the word *kabbalah* have the meaning "that which is passed on."

The *Sepher Yetzirah, The Book of Formation*, ascribed to Rabbi Akiba (circa 120 CE), is a root text of what can be written. Even this is so venerated that some Kabbalists give its authorship to Abraham himself. An even loftier provenance (according to the account of Manley P. Hall's *The Secret Teachings of All Ages*) was enshrined in the belief that the *Sepher Yetzirah* was first "spoken to the angels" by the Ineffable, before the creation of the world. This version has it that from Adam, who had it from the angel Raziel, it was progressively passed down by angels into the time of King David. After an influential edition of it in the sixth century CE, creditable historical sources have the *Sepher Yetzirah* finding its way to Moses de León, the writer of the *Zohar* (*The Book of Splendors*), in around 1350.

The *Sepher Yetzirah* approaches "the two wisest things" of the Pythagoreans—numbers and naming—in terms of three principles (*seraphim*): numbers, letters, and vocals (sounds). It then names ten further properties (*sephiroth*) and moves on to the essential structure of the Hebrew alphabetic language itself.

There are three "mother letters," seven "double letters," and twelve "simple letters"—twenty-two in all (shown in fig. 15.6, p. 164). These emanate from the ten properties of the Ineffable. The number 32 is sacred as the number of this entire group. The ten properties are a cycle, described as "having its end joined to its beginning." The cycle of the ten became, in medieval times, the diagram of the Tree of Life symbol, and all Kabbalistic streams seem to agree that the top three *sephiroth* (spheres) are the essential dynamic of the united whole. Pythagoreans, working from the Tetraktys, conclude similarly (see figs. 15.7 and 15.8).

One sees, reading from the top, that the Tetraktys establishes a triad as its matrix. The further seven are all integral to the pattern of ten. The same thought, geometrically speaking, is embodied in the triad crowning the Tree of the *sephiroth*.

The relationship of the 3 and the 7 is centrally present in the areas outlined by the Starcut's first mid-side diagonals. In figure 15.3 (p. 160), the central triangle is of area 6, and the surroundings total to an area of 14. There is a 3:7 proportion, as in the two sacred forms pictured in figures 15.7 and 15.8.

I find myself increasingly convinced of a connection between the Kabbalah and the Starcut diagram. There is considerable metaphorical strength about the association of the central triangle's area with the D'BIR cube number of 216—Pythagoras's perfect cube. The triangle's total edge length, its interface, measuring 72, tallies so well with the concept as well as the detail of the Holy Names used for harmonizing with the Divine Essence through sound, and with the tradition that the seventy-two words are composed of 216 letters—as was written by Rabbi Todros ben Joseph (1879) quoting thirteenth-century Kabbalistic literature: "It is known to masters of Kabbalah that seventy-two names surround the seat of glory."

The 3 and 7 ratio as a characteristic of the 10 is also convincing.

The translation from the Bible's public written measures into the hidden geometry is achieved in the Bible's 8,000 cubic volume having a face of 2,400—in the same 6:20 ratio that exists between the whole square and its central 3-4-5 triangle, which also can be interpreted as a cube.

There are other interesting correspondences: the number 144; the fact that there are thirty-two 3-4-5 triangles in the diagram's eight-pointed star; and that there are thirty-two "voices" that comprise all the letters and numbers in the *Sepher Yetzirah*. Some Kabbalists refer

Figure 15.6

In Hebrew, there are three "mother letters," seven "double letters," and twelve "simple letters"—twenty-two in all.

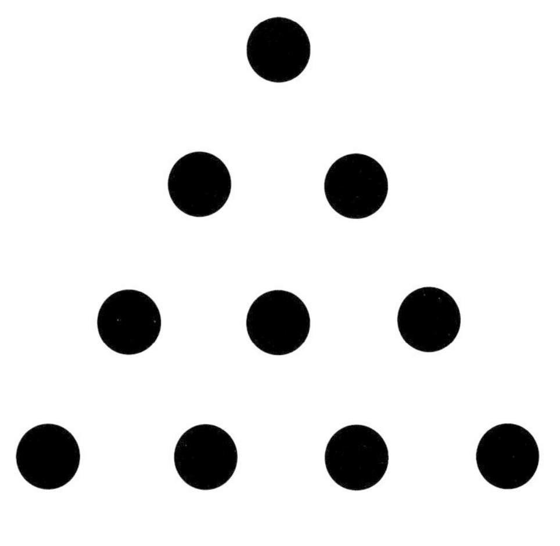

Figure 15.7

The triangular matrix of the Tetraktys of the Pythagoreans.

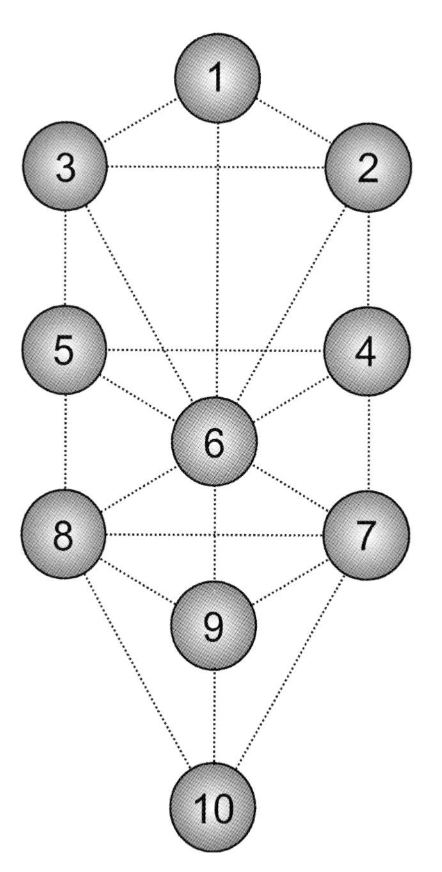

Figure 15.8

The cycle of the ten properties of the Ineffable became, in medieval times, the diagram of the Tree of Life symbol, with the top triad functioning as the essential matrix of the whole.

to the thirty-two as the "Teeth of the Great Countenance" (fig. 15.9). I enjoy thinking of the star form this way!

Overall, this star's area is 432, as we saw in chapter 14; it is a number we will meet again.

The entire square is triangulated by forty triangles, and the number 40 certainly has public biblical credentials in terms of forty days of the Flood and the forty years the people spent in the wilderness between release from bondage and "homecoming." Moses, according to the Kabbalah, spent three sets of forty days on Mount Sinai. During the first forty days he received the public Law of the people; during the second forty days he received the Mishna, the law as understood by the rabbis; during the third period of forty days he received the Kabbalah, the teaching as transmitted to the specially prepared. There is an overarching significance of 40 in the synthesis of Kabbalah where 40 is the master number of the four Trees, each with ten *sephiroth*, pervading four orders of Creation (see fig. 15.10). It seems appropriate that 40 represents the complete triangulation of the Starcut square.

Figure 15.9

"The Teeth of the Great Countenance."

Before leaving the topic of Starcut/Kabbalah correspondences, two further associations with the number 72 are worth mentioning, remembering that 72 is the boundary of the 216 triangle and that 720 is the area of the whole square. The sacred four-letter name, not to be spoken at all by religious Jews, is known as the Tetragrammaton: יהוה (YHVH). Its value by gematria is 26. Sometimes, however, it is deployed in the form of a Tetraktys, whereby it is transformed into the 72 as in figure 15.11.

The number 720, in ancient measures, is not just any old number. It is a consummate sexagesimal number since it is 12 × 60. The 3, 4, 5 series adds to 12 and multiplies to 60, and these are even now the governors of day and night time. The number 720 is 1 × 2 × 3 × 4 × 5 × 6—factorial 6—written by mathematicians as 6! It is also 8 × 9 × 10. And since, in the numerology of the Kabbalah, any number multiplied by 10 is considered to be the full manifestation of whatever principle that number embodies, 720 may be taken to be 72 amplified into the square of manifestation—Zahir, the Manifest.

So what is this all about? It is said that ancient Hebrew, like ancient Greek and Sanskrit, was a hieratic language, meaning it was tailored to the priestly system of knowledge. The Kabbalah believes itself to be the gift of a higher aspect of mind, and so uses the word *revealed* to specify its own arising. Did this revelation actually arise in the contemplation of laws of number that could be recognized in the rationale of primary geometric figures? Are the Starcut square and its numbers an ancient musing board that catalyzed some very early number theory, geometric knowledge, and also metaphysical naming?

We know so little about what abstract tools may have facilitated learning at remote stages in history. We see results. We have Euclid, and the classical mathematicians and philosophers; they tell us of conclu-

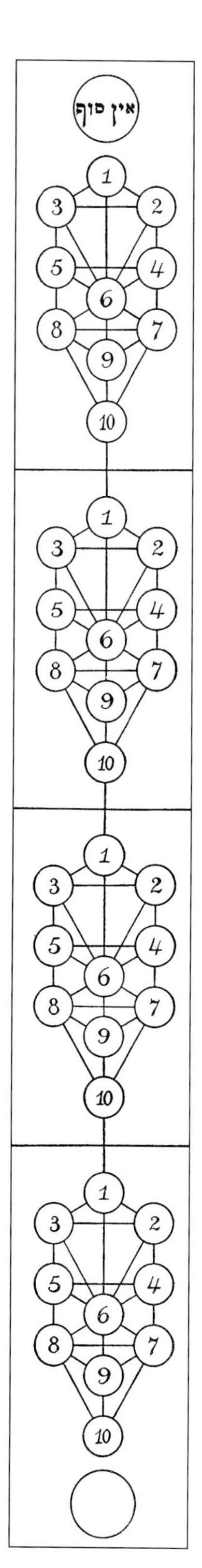

Figure 15.10

Forty is the master number of the four Trees, each with ten *sephiroth*, pervading four orders of Creation. Forty also represents the complete triangulation of the Starcut.

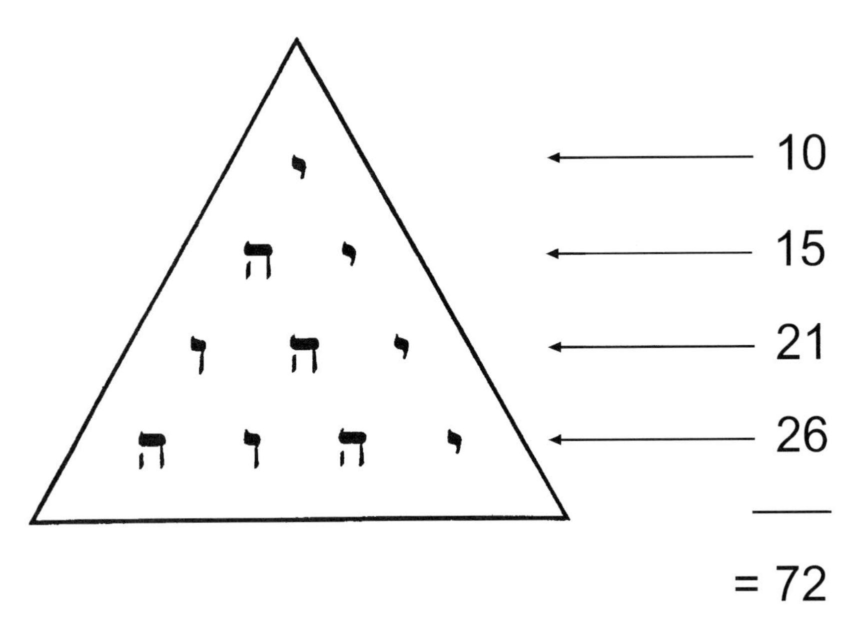

Figure 15.11
The Tetragrammaton in the form of a Tetraktys, whereby it has a value of 72.

sions reached over the millennia before them. But what were the devices that stimulated the creative processes that were already going on when various cultural strands first came into sight over the horizon from prehistory? Could one of them have been this simple—seminal—geometric star of triangles inside a square, drawn in the sand?

Words from a dissertation of Christopher Powell, a student of Mayan history, though used about a different culture, apply equally here:

> All that is left to this age are the fragmentary end products of the application of methods and theories developed and used for . . . millennia by societies, that in many respects, remain mysterious, or at least inadequately understood. Those with an interest in this field are left to absorb what data is available, deduce what the ancient problems were, and then to decipher what methods may have been used to solve them.
>
> C. Powell, "A New View on Maya Astronomy," M.A. Thesis, University of Texas, 1998

A further fascinating hint about the early choice of a number system that we have inherited, but for which we really have no conclusive explanation, emerges when we contemplate the relationship of the Starcut diagram with its related circles. We shall see this in the next chapter.

Opposite: Figure 15.12

Nearly . . . It is often not appreciated that, in strict geometric terms, the 360° circle cannot be drawn without approximation. The smallest whole-number arc that can be drawn with straightedge and compasses only is 3° (shown in red). So in whole-number geometry, the 120° division of a circle is the finest possible. The common protractor, for all that it can do in practice as a precision instrument, is not geometrically "pure." Its construction actually depended on trial and error in order to trisect the 3° segment. And why 360°? It is divided by all single integers except 7, and it loosely approximates to the number of days in the year—but could there have been a clincher that made its adoption irresistible?

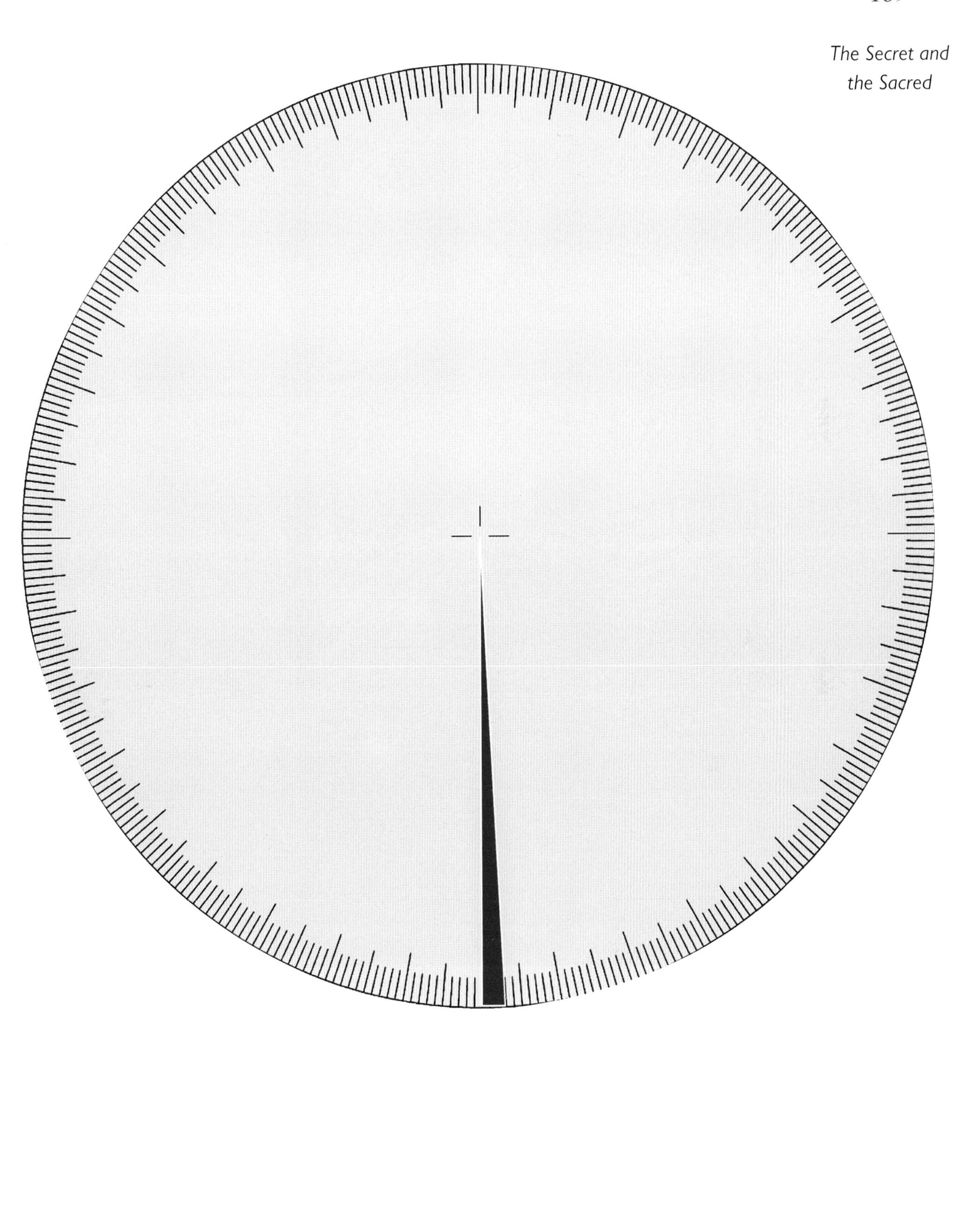

Figure 16.1

Two spiral galaxies in an early stage of merging, located in the constellation of Cetus, the Whale, about 350 million light-years from Earth, taken by the Hubble Space Telescope in 2008.

16
Wheels of Number and Time

It is not necessary to have any numerical equivalent for π (pi) to recognize that there exists some constant that, when multiplied by a circle's measurable dimensions, allows one to name and utilize one's results. Expert metrologists have identified several ancient approximations for π; my simple purposes only require that my "sand reckoner" saw that such a constant existed, not that he or she could define it with any numerical accuracy.

Remembering that the number system is self-generated by the diagram, it seems to be whispering something. The two most obvious examples of circles, inner and outer, suggest 360 and 36. In circle form, the sexagesimal continues.

The perfect number series characteristic of the 3-4-5 triangles is reflected in the circles within those triangles, the smallest of which has an area of 1π, since in a 3-4-5 triangle the radius of the in-circle is 1.

There is a theorem that states that the circumference ($\pi \times$ diameter) of a circle is equal in length to the circumferences of any number of circles inside it, if their diameters add to the same length as the big circle's diameter. This is shown elegantly in the figure that uses the circumcircles of the 3-4-5 triangle series (16.4, p. 173). Three of the hypotenuses sum to the length of the fourth.

One application of the Starcut to circles solves the comparative sizes of circles in square clusters (see fig. 16.5). It turns out that the 7 × 7 checker is the key. The sloping yellow lines, passing through Starcut nodes, give seventh cuts of the square.

In *A Little Book of Coincidence,* John Martineau shows that the circles in this cluster are minutely close to the relative sizes of Jupiter and Mars. What the Starcut adds is a way to analyze these relative sizes in a simple geometry, just as its eleventh cut did in connection with the comparative sizes of the earth and the moon (in chapter 6). On an occasion when I remarked about such seemingly remarkable coincidences, another friend, Giles Peppe, pointed out that in a relatively stable environment such as our solar system, one may expect to find that the attractors naturally have a harmonic order, otherwise the system would become chaotic and self-destruct. It is, however, surprising just how geometrically simple some of the outcomes are.

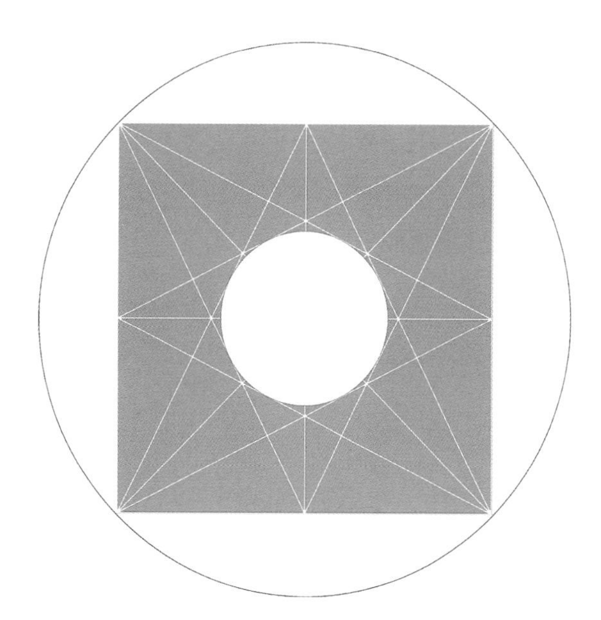

Figure 16.2
An outer circle (circumcircle) around a square of area 720 has an area of 360π. The circle within the central octagon, also within the big 3-4-5 triangle (its in-circle) has an area of 36π.

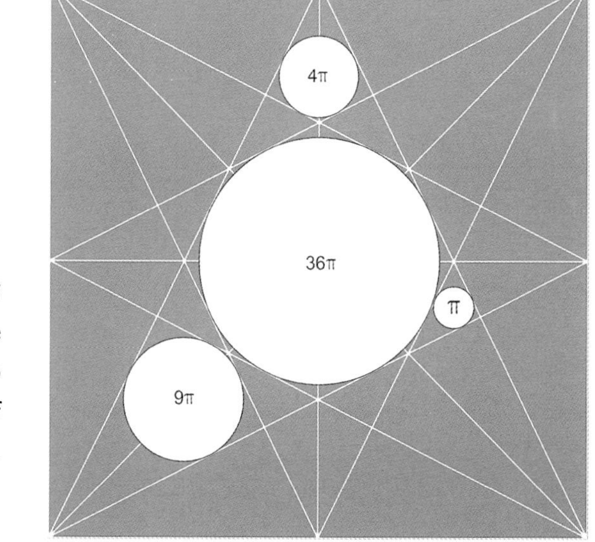

Figure 16.3
The series 1, 4, 9, 36 is present as in-circle areas in the four versions of the 3:4:5 ratio triangle. Recall that these are the squares of the first perfect-number series.

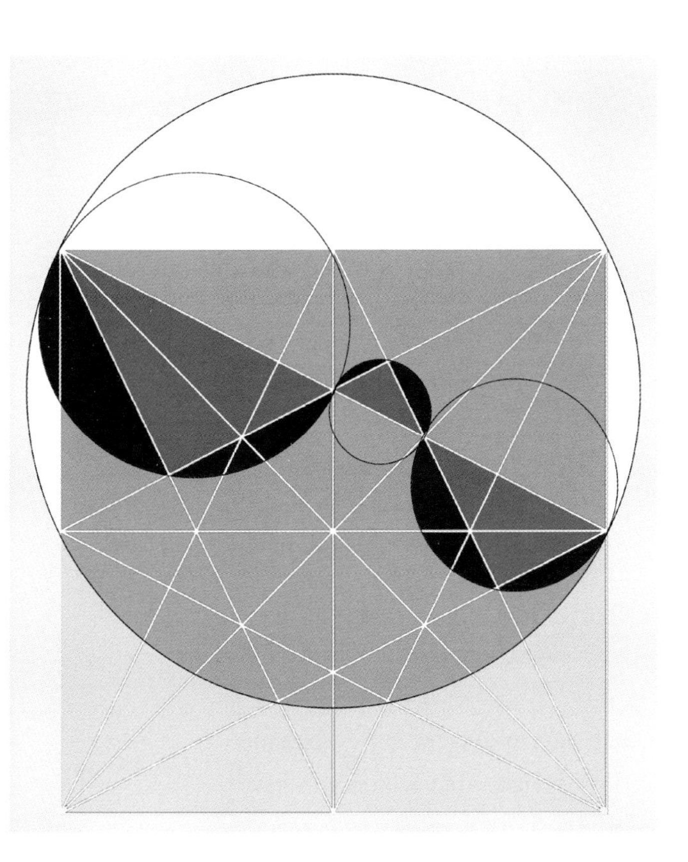

Figure 16.4
This figure elegantly shows that the sum of the hypotenuse-diameters of the three smaller 3-4-5 circumcircles is equal to the diameter of the larger 3-4-5 circumcircle, which in turn demonstrates the theorem that the circumference of a circle is equal in length to the circumferences of any number of circles inside it, if their diameters add to the same length as the big circle's diameter.

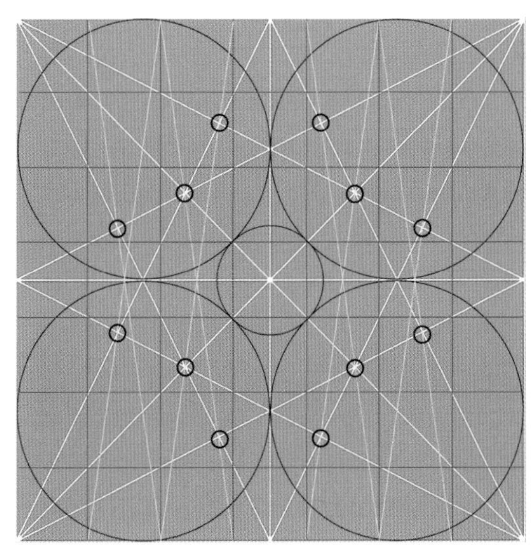

Figure 16.5
The small circle has a radius that is $^1/_{14}$ of the square's diagonal. The diameter of the small circle is $^2/_5$ of the diameter of the big circle.

Further numerical aspects of the Starcut appear when its component forms are rotated in a graphic multiplication table. Starting in the top left corner in figure 16.6, going clockwise, the area numbers of the big 3-4-5 triangles and the bordering 1-2-√5 triangles multiply 8 times. The 90° right angles do likewise.

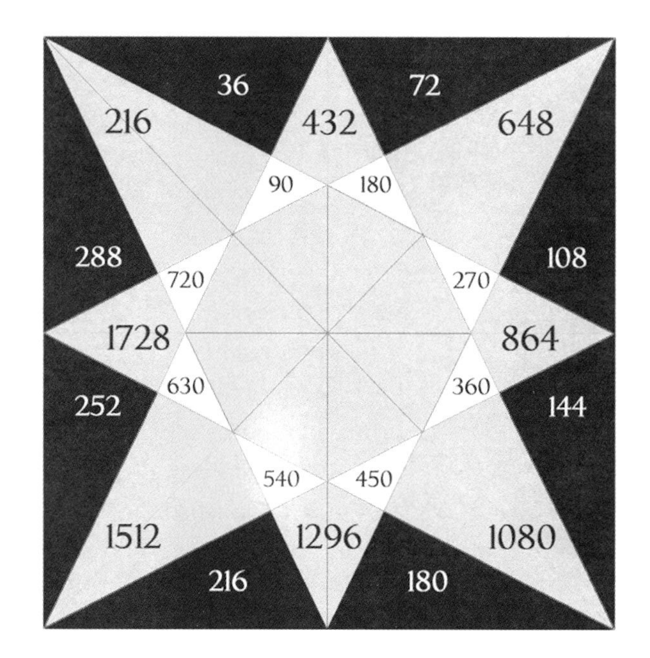

Figure 16.6
The Starcut can also function as a multiplication table: moving clockwise from the top left, the area numbers of four sets of triangles—including the big 3-4-5 triangles, and the bordering 1-2-√5 triangles—are multiplied 8 times: by 1, by 2, by 3, and so on.

If you were an astronomer/mathematician in ancient times, musing on the numbers and their multiples in this device, what would you have made of the coincidence that the rising of the equinoctial sun moves one degree against the backdrop of stars every 72 years (an observation that could easily be made); that the great precessional wheel of the sky enters a new zodiacal sign every 2,160 years (a calculation that could easily be made)? Might you have noticed that your system of weights and measures incorporated the same numbers—the moon, for instance, being accepted as having a diameter of 2,160 "miles" (for which you would have had some other word)? Would you have reached the conclusion that the numbers suggested by this simple form, which accorded with measures and cycles in the heavens, must in some way be sacred? Especially since the whole construction had started from two lines that revealed themselves to have the value of the fingers on each of your hands. Would you have noticed that all the numbers in the diagram—when their component integers are added to each other—always add to 9? That the number 9 (as the Vedic mathematicians concluded) is the number of completion and that 90 is perfect as the number of a right angle, fitting fourfold and exact into the circle of 360°? However, perhaps none of this is coincidence—perhaps the number systems came from this device in the first place. (The next chapter explores one of the numbers in the pattern above—108—as an example of how extensive the correlations of number, geometry, and cosmology can be, and how this harmony has been reflected in spiritual practices.)

Below is a section from an address given by the notable scholar Sir William Jones, the founder of comparative philology, who was the first president of the Asiatic Researches Society, a learned association that met regularly in Bengal in the late 1700s, and from whose transactions the passage is excerpted. It was a time when Hindu pandits clearly respected the interest of some of their British enquirers and were willing to confide Brahmanic cosmological material in fresher mint form than the summary listings that we now have from that tradition.

Sir William, in a public address on Hindu chronology, is discussing interpretations of the numbers involved. Here I am not seeking to elucidate his (occasionally opaque) argument, but simply to give an example of the numbers that are to be met with in this ancient system:

> The 4,320,000 years of which the four Indian ages are supposed to consist, mean [according to an earlier argument] only years of 12 days, and in fact that sum, divided by 30, is reduced to 144,000: now 1,440 years are one pada, a period in the Hindu astronomy: and that sum multiplied by 18 amounts precisely to 25,920, the number of years in which the fixed stars appear to perform their long revolution eastward. The last mentioned sum is the product of 144 which was an old Indian cycle, into 180, or the Tartarian period called Van, and of 2,880 into 9 which is not only one of the lunar cycles, but considered by the Hindus a mysterious number and an emblem of Divinity, because if it be multiplied by any other whole number, the sum of the numbers in the different products remain always 9, as the Deity who appears in many forms, continues One immutable essence. The important period of 25,920 years is well known to arise from the multiplication of 360 by 72, the number of years in which a fixed star seems to move through a degree of the great circle. . . . We need only compare the two periods 4,320,000 and 25,920 and we shall find that, among their common divisors are 6, 9, 12, 15, 18, 24, 36, 72, 108, 144, etc.; which numbers with their several multiples, especially in a decuple progression [that is,

> multiplied by 10s], constitute some of the most celebrated periods of the Chaldeans, Greeks, Tartars, and indeed of the Indians. We cannot fail to observe that the number 432, which appears to be the basis of the Indian system, is a sixtieth part of 25,920, and by continuing the comparison we might probably solve the whole enigma.

All the numbers above, in multiple and decuple form, emerge directly from the sexagesimal system intrinsic to the Starcut diagram. I recently came across a student of ancient culture on the internet scratching his head as to why on earth the standard weight/volume of the Sumerian *talent* should have been set at 933,120 grains of wheat, which it was. Well, for a start it is 2,160 × 432, or 4,320 × 216. And it seems a fascinating coincidence that if the great 216 triangle is rotated through all the 120 divisions of the circle that can be drawn by geometry alone—the number 120 also being the area of the central octagon of the whole figure—the result is 25,920, the traditional number of the complete precession of equinoxes.

While reflecting on the way the Starcut numbers coincide with Eastern numerical traditions, it is worth mentioning the correspondence with that of China. We will return to the Chinese connection at the end of this book, but in the present context the following quotation is irresistible. It is from the classical author Ts'ai Yung writing about the perfect conceptual measures for the Ming Tang, the "Luminous Hall"—an idealized architectural microcosm wherein the Imperial Palace on earth mirrored the perfection of the heavens:

> The base of the hall is a square of 144 feet on each side, this being the number of the trigram k'un [earth]. The roof is circular, of diameter, 216 feet, which is the number of the trigram ch'ien [heaven]. The Great Ancestral Temple of the Luminous Hall is a square with sides of 60 feet and the Chamber for Heavenly Communication is 90 feet in diameter. . . . It has 9 rooms symbolizing the 9 provinces. It has 12 rooms in accord with the 12 hours of the day. It has 36 doors and 72 windows. The doors all open to the outer world and are not shut, thus showing that throughout the realm nothing is concealed.

Figure 16.7
The first trigram: *ch'ien*—heaven. Its number is 216 in Chinese numerology.

THE GREAT SQUARE—72 AND 720

The sequences in the multiplication graphic in figure 16.6 (p. 174) are blazingly relevant to the numbers that occur both in the practical and the visionary measures of antiquity. The huge topic of ancient metrology belongs to specialists who have studied it far more deeply than I. After my first public lecture on the Starcut, the late John Michell cornered me. Referring to his work on the ratio between the sizes of the earth and the moon, that I had shown could be derived directly from the square, and to the numbers that are automatically established within the Starcut, his first question was: "Is the geometry absolutely accurate?" On being assured that it was, he insisted that I should publish. It is to books such as his that the reader who wishes really to grapple with the *Dimensions of Paradise* (one of his titles) should turn. Here is a quotation from its chapter entitled "The Numbers of the Canon":

> In the operations of simple arithmetic and throughout all the numerical operations of nature, such as the periods and intervals of the solar system, certain "nodal" numbers occur, providing a link between processes and phenomena which otherwise appear quite unconnected with each other. Most prominent among these are multiples of 72.

Then, starting with 720, he lists thirty-six such numbers and goes on to explain how this master number was also to be found in the gematria of the names of gods, and in the symbolic measures described in contexts such as Plato's perfect city in the *Laws,* in monumental structures such as Stonehenge, and in the dimensions of the visionary New Jerusalem.

To this much can be added. The number 720 is the fundamental multiple of the spherics. The tetrahedron contains 720°; the octahedron includes 2 × 720°; the cube has 4 × 720°; the icosahedron has 5 × 720°; and the dodecahedron has 9 × 720°.

The central angle of the radials of a pentagon is 72°, and the number turns up all over the world, often in connection with complete sets or groupings. The number seems to be used intentionally this way, as can be seen when its use is actually inaccurate, such as in the listing, in Genesis, chapter 10, of the 72 races descended from Noah, where the descendants of Peleg are not counted because the number would be exceeded. There were said to have been 72 Jewish elders associated with the compilation of the biblical text known as the Septuagint. In Moses de León's *Zohar,* the great classic of Kabbalah, there are 72 elders of the synagogue and 72 steps on Jacob's ladder. There are 72 immortals according to Taoist tradition, and Confucius was said to have had 72 disciples. Is it coincidence that 72 beats per minute is the accepted average pulse of a human being?

As a further reflection upon significant numbers that occur in the Starcut, the next chapter contains a few more of John Michell's findings. It is an adaptation of an article I wrote some years ago, commemorating the occasion of the 108th edition of Johann Quanier's *New Humanity* magazine. It may provide a summary example of some of the themes that have threaded their way through these pages.

Opposite: Figure 16.8

This is a page from Athanasius Kircher's book *Oedipus Aegyptiacus,* published in the mid-seventeenth century, a massive compendium of Egyptian, Chaldean, classical, Gnostic, and Hermetic lore. It shows the name of God translated into 72 languages and, above the mandala-like central device, 72 leafy fronds are labeled with the divine qualities of the Kabbalah. Kircher was a Jesuit, and his book was published in Rome at the height of the Counter-Reformation, so it is no surprise that in the middle of this huge theological confection sits the figure of Christ in majesty.

Iconismus inserendus tom. II Fol. 287
Speculum Cabalæ mysticæ
Inquo omnia, quæ Hebræi de nomine
dei tetragrammato arcanè retulerunt,
ys ad nomen messiæ Iesu respexisse de
monstratur Omnes quoq. mundi nationes
nomen dei non sine mysterio
4. litteris enunciare, docetur.
Explicatur in Cabala fol. 287
Arbor Cabalistica
oculus sinister Gabriel
oculus dexter Raphael
Auris sinistra Michael
Auris dextra Samael
Raph. ziel
Ephraim
Gad
Manasses
Iudas
Isaschar
Nephtali
Dan
Aser
Beniamin
Simeon
Ruben
Zabulon

Figure 17.1

The 108 mala beads are used to count the repetitions of mantras or the names of God. The number 108 is a multiple of 9.

17
Powers of 108

In the commemoration of everyday events the number 21 represents a coming of age, and a centenary is the completing of a great cycle; in a more esoteric commemorative usage the number 33 might be the coming of age and the great completion perhaps denoted by 108. Here are a cluster of associations concerning this latter intriguing number that is one quadrant of the star form's area of 432.

At the outset I must admit to some favoritism, having been for many years a member of Oscar Ichazo's Arica School of philosophy wherein 108 characteristically governs the practice of mantramic repetition. It was this that first drew my attention to this number. The "Apollonian" or Pythagorean tradition sees certain numbers as being as much qualitative as quantitative. Augustine of Hippo, whose Christian spin tended to distort his classical sources, did, however, speak authentically for tradition when he wrote:

> Numbers are the thoughts of God . . . the Divine Wisdom is reflected in the numbers impressed on all things . . . the construction of the physical and moral world alike is based on eternal numbers.

The first indication of the quality of a number comes from "casting out 9s," that is, compressing it to a single integer by adding its components. So 108 is therefore 1 + 0 + 8 = 9. Nine is the culmination

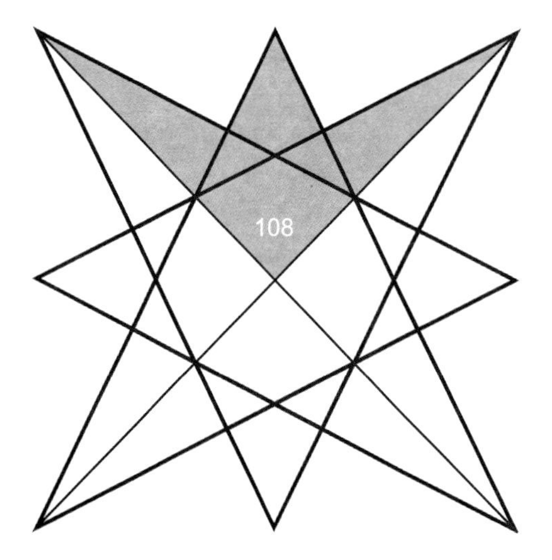

Figure 17.2
The number 108 is one quadrant of the star form's total area of 432.

of the basic series and the herald of a new series at ten. So we might expect to find some wider associations with 108, and we do—in abundance. Sufis say that numbers go on unendingly, and numbers summing to 9 represent aspects of that infinity.

In India, China, Tibet, and central Asia, rosaries are used to count the repetitions of mantras or the names of God. Such rosaries traditionally have 108 beads. The spiritual practice of repetition has many layers of significance. One such layer is that with each bead there is a pulse of thought and sound (the sound may be vocalized either externally or internally) that is a "moment" wherein is the coming-to-be, the clarification and the passing away of a complete universe. The "eternal now" thread upon which all universes are strung is represented by the string of the rosary; and the Sanskrit name for a thread is *sutra,* a word appropriately applied to the verses of the Buddha's teaching.

The number 108 equally occurs as the cycle-count in Chinese Taoist disciplines where the practitioner performs certain breathing and visualization techniques in multiples of this number. In the Chinese "Pure Land" School of Buddhism, 108 and its "amplitudes" (for example, 1,080, 10,800, and so on) are a constant in the numbering of spiritual entities, states of being, and blessings. The number of zeros on the end of a number might be considered as symbolizing the manifestation, intensification, and exaltation of the quality already there. So 108,000 is, as it were, the number 108 with its volume turned up very loud!

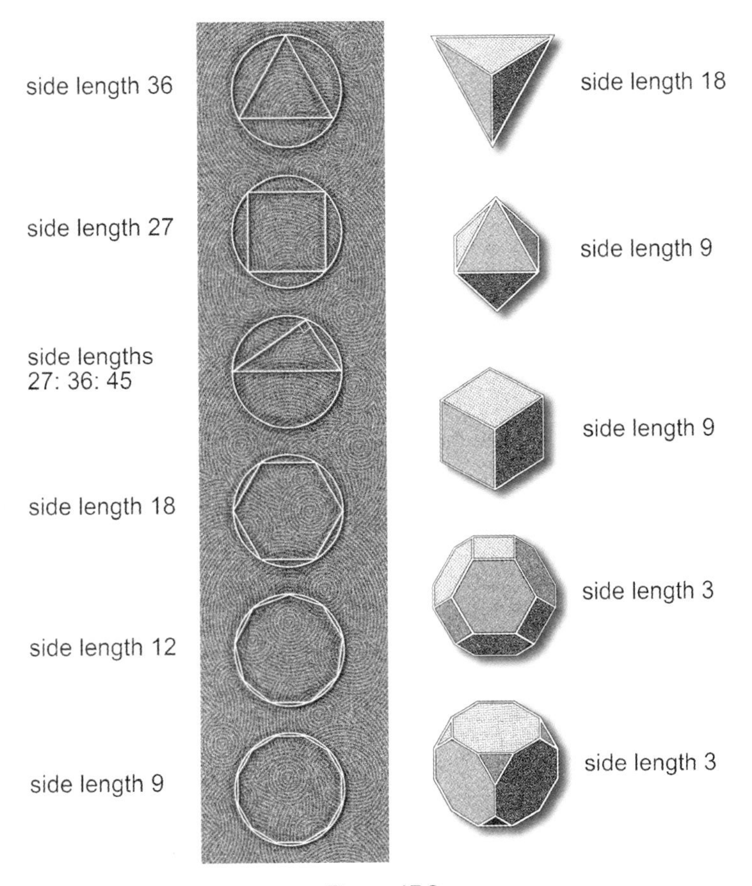

Figure 17.3

The number 108 is very versatile as a perimeter made up of the sum of the sides of geometric figures.

The Rig Veda, humanity's oldest extant scripture, contains 10,800 stanzas. The traditional Hindu cosmology considered 108,000 years to be one season or quadrant of the Kali Yuga, the last of the four ages of cosmic time, and the one in which, according to this tradition, we are now living; and 1,080,000 years is one quadrant within the 4,320,000 years of the total cycle comprising all four yugas.

In the Greco-Chaldean astrology of the West (and, until quite recently, the astronomy too) the measure of 25,920 years has been taken to be the time that it takes the sun, which does not quite complete a zodiacal circuit each year, to drift slowly backward through all twelve zodiacal signs. This "precession of the equinoxes" is sometimes referred to as a "heavenly day" and one hour of that day (1/24) lasts 1,080 years.

Hipparchus, the astronomer credited (many believe, wrongly) with the discovery of the "precession," also claimed that there were only 1,080 first-magnitude stars in the sky. Earlier, Heraclitus the Ionian had said that, in the inexorable wheel of change, the fire, which was for him the fundamental element of cosmos, would consume one age of civilization every 10,800 years. Why? What is it about 108 that carries this quality of a completed cycle?

Its most rudimentary components are particular forms of 1, 2, and 3—each of them carrying its own mathematical "power": $1^1 \times 2^2 \times 3^3 = 108$.

In the Pythagorean school of thought it was said that with 1, 2, and 3 all number is implied. In the next chapters on the Tetraktys we will see this illustrated. Here 1 is, in a qualitative sense, not a number at all, since it denotes the Unity that cannot be divided or the essential individuality that cannot be multiplied. The first number was 2; though some said that the "even" 2 was not properly a number since it is, qualitatively, a principle establishing both "other and same." Whatever one's view of the 2, however, the 3 fully establishes number; and from the 2 and the 3 within the 1, flowed all of creation.

So 2, given its own power, that is, squared, with 3, given its own power, that is, cubed, produce 108. The squaring and cubing are not incidental arbitrary operations. When we square a number it becomes

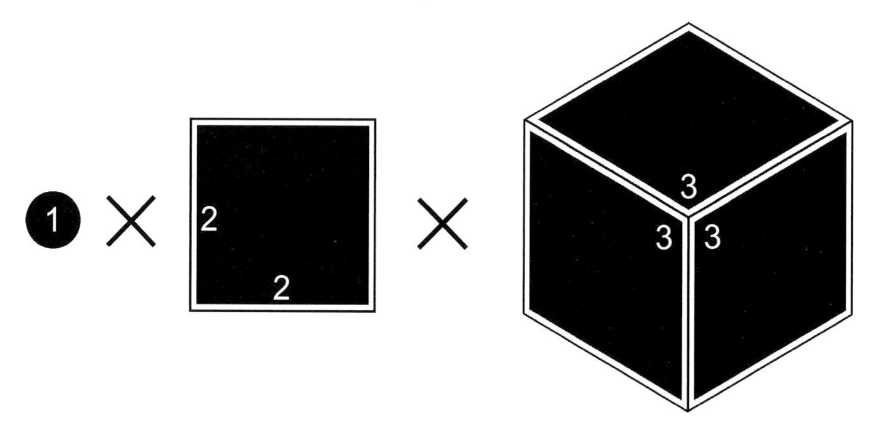

Figure 17.4

The most rudimentary components of 108 are illustrated here: $1^1 \times 2^2 \times 3^3 = 108$.

an area; when we cube it, it becomes a volume. Our three-dimensional cosmos is therefore directly there in the first numbers; and the relationship back to Unity is further affirmed when we recall that the $\sqrt{2}$ is the length of the diagonal of a unit square, and that the $\sqrt{3}$ is the diagonal of a unit cube. So 108 is, qualitatively speaking, intrinsically geometric. Figure 17.3 indicates that very clearly.

The number 108 is also 9 × 12. And 9 is the number of completed structure, a triad of triads, or the pattern of creative process working upon itself, 3 always being associated with the basic dynamic—hence the trinities of creation theologies. On the other hand, 12 is the number of time and frequency. It sections the circle through natural twofold and threefold operations (hence our clock face). The number 12 is also integral to all the spherics.

The tetrahedron in the center of figure 17.5 has 12 angles (of 60°, hence its 720° total); the octahedron and the cube both in the lower part of the figure have 12 edges; the icosahedron has 12 apices; and the dodecahedron has twelve faces. The number 12 governs the chromatic

Figure 17.5
Platonic solids all having the number 12 governing some aspect of their construction.

scale of music, there being 12 distinct musical pitches defined by the cycle of fifths—a topic to which we shall return.

There has already been reference made to the coding of numbers and letters in classical Greek. John Michell (following the archaeologist Frederick Bligh Bond) pointed out that the numerical value of the Greek New Testament term for the Holy Spirit (το πνευμα αγιον) was 108.

So in the rhythms of sacred repetition, in cycles of time and space from the Orient to the West, and from the Vedas to Christian scripture and through to current spiritual practice, 108 constantly features.

It is also there in the sun, the moon, and the stars.

If you multiply the diameter of the sun by 108, you get the mean distance of the sun to the center of the earth. The radius of the moon is 1,080 miles (the mile, incidentally, being an ancient measure stoutly defended by many for reasons other than mere British chauvinism). And, by odd coincidence, the metal associated with the moon—silver—has an atomic weight of 108. And as for the stars—see the star in figure 17.6.

The shape of the Great Pyramid (indeed any four-sided pyramid) has a total of 1,080° in its four face triangles and its square base. The number 1,080 is also the total of degrees of angle around the inner perimeter of an octagon—a symbol of regeneration.

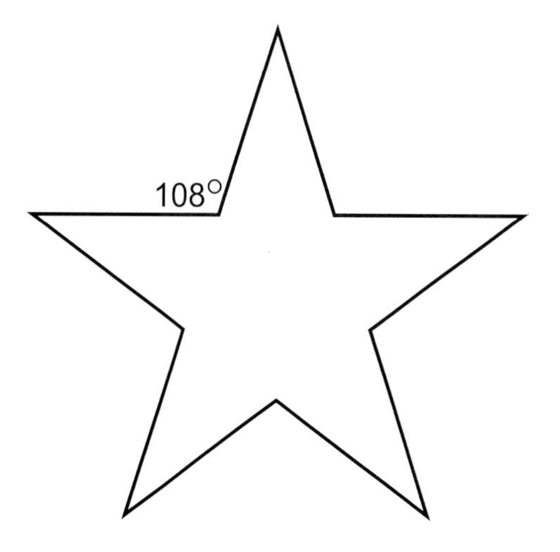

Figure 17.6
There are five 108° angles in the pentagram star.

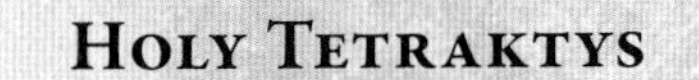

Holy Tetraktys

Bless us Divine Number
Generating gods and men
Most holy tetraktys
Containing root and source
Of ever-flowing Creation.

For divine number begins
In profound pure unity
And coming to the holy four
Begets the mother of all.

The all encompassing
The firstborn
Never swerving nor tiring
The holy ten
Keyholder of all.

Figure 18.1

The horizon of the ineffable: According to Iamblichus, Pythagoras taught his lore from a triangular arrangement of ten pebbles. It is known as the Tetraktys ("the fourfold figure"). In addition to its strictly numerical aspects, the Tetraktys can function as a "symbolic abacus" suggesting analogies for concepts that Plato would have called "Universals." The Pythagoreans took an oath invoking the blessing on page 187. The Tetraktys is Pythagoras's own basic map of how reality is constituted.

18
Ten Pebbles in the Sand

We have a most interesting Pythagorean idea to contend with—that ideal number is not necessarily subject to a sequential or causal progression from one through ten, but is rather a unity with ten essential and potential qualities, simultaneously present in the Decad or Tetraktys.

KEITH CRITCHLOW, INTRODUCTION TO
THE THEOLOGY OF ARITHMETIC

The senior cosmogram of the Hellenic numerical tradition was the Tetraktys. It was a geometric/numerical figure revered in the schools of Pythagoras and Plato and by a good number of those who have followed their metaphysical lead. It was considered to express the fundamental emanation of the principles that originate and sustain all of manifest reality, not just at a putative past "moment of creation," but right now.

The Tetraktys could refer to a primacy of principles present in every moment. That means in the perceptual/thinking/reacting core of all human experience. Conceptually it reaches deep into the mind's knowledge of causes. Visually the eye moves, recognizing patterns of points; mentally the thought constructions easily become numerical (figs. 18.2–18.4).

Figure 18.2

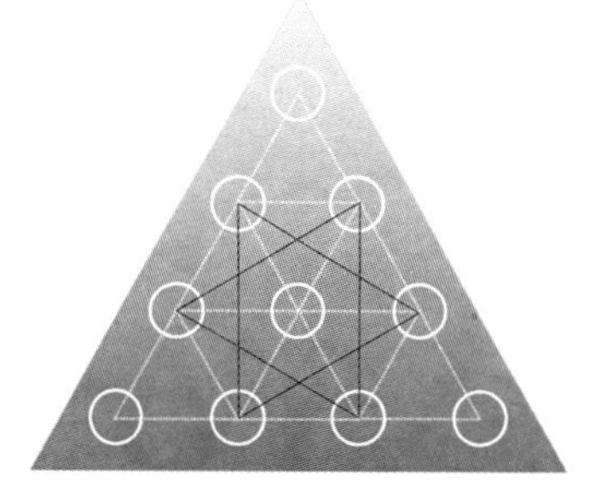

Figure 18.3

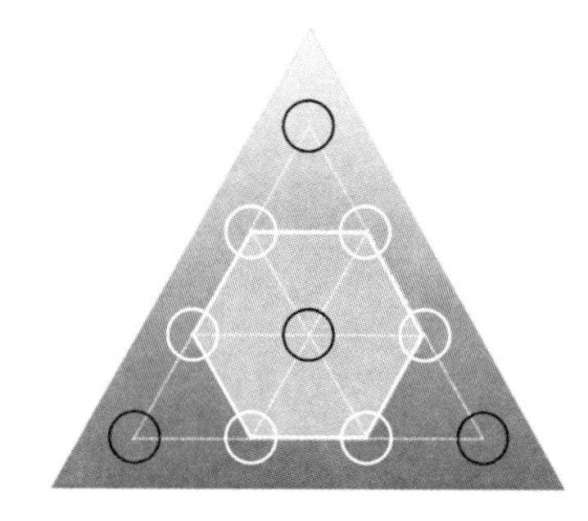

Figure 18.4

Figures 18.2–18.4. Patterns of points in the Tetraktys.

Meanwhile the concept is in essence a unity. In the Pythagorean understanding the whole thing is present first as the creative Unity in act and second as a description of emanation in causal space and time. As such the Tetraktys is a metaphor, another mnemonic even, denoting a concept on the very brink of the unknowable. No wonder it was honored with an oath. It was said that Pythagoras taught all his lore from this array. For those interested in its many applications, K. S. Guthrie's *The Pythagorean Sourcebook* lists eleven of them as given by Theon of Smyrna.

As a designer, it was the geometry that first spoke to me.

As shown in figure 18.5, the geometry comes directly from a sixfold vesica upon a circle of origin. Six-pointed stars suggest yin-yang-type dynamics in the figure.

However, the foremost numerical impression of the Tetraktys is its "1-2-3-4-ness" that constitutes its "10-ness." In this metaphysical metaphor, 10 is the full amplitude (the volume) of the primal utterance of "Being" itself—the 1. Ten is one fully expressed in terms of becoming active. So in the 10 lie the principles of creation. This elaboration from the 10 is analogous to what happens with the Starcut where the characteristic mid-side diagonals establish a relationship involving 1, 2, 3, and 4 as sections of two lines each of length 5. As we see in figure 18.7 (p. 192), there is a substantive similarity though the information is differently deployed.

In some respects I first saw the sand reckoner's square as the fuller manifestation of the bottom (four-point) line of the Tetraktys; but learned from Professor Critchlow that a set of numbers known as the *lambda* could be substituted for the simple pebbles, or points, of the

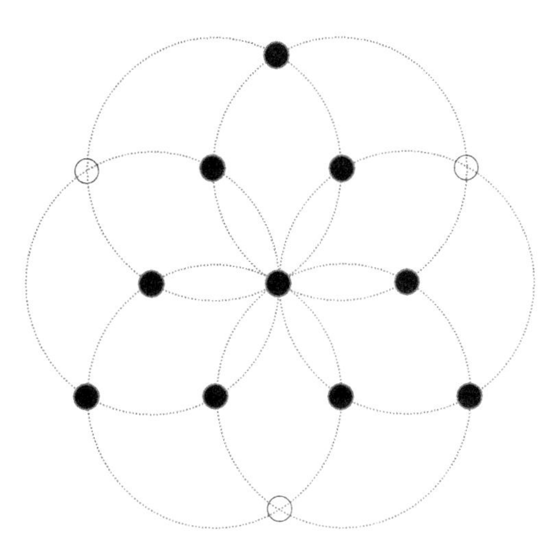

Figure 18.5
Ten points are selected from the cluster of thirteen points that is made by the overlapping of seven circles in multi-vesica form. Seven circles, in a different format, generate the Starcut.

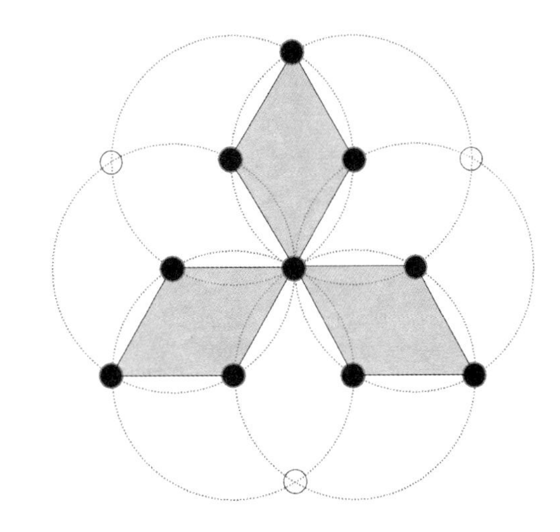

Figure 18.6
The shaded areas are three of seven rhombuses that can be found in the geometric arrangement of the Tetraktys.

array. It was this piece of information that provided the basis of my own work on it, which, however, needed a further catalyst that came from the philosophy of Oscar Ichazo.

Since Ichazo's name has come up before and is unfamiliar to many, I would say only that he is a contemporary philosopher who has founded a school (Arica) through which the ancient "path of return"—the quest of all schools of spirituality—is offered in contemporary terms. Since his work proposes a complete map of the psyche, from dense subjectivity right through to the highest transcendent states of consciousness, it is an immense body of practice and theory. There is no concern here to persuade the reader of his propositions but rather to show the context I started with in what follows.

An early account of part of Ichazo's work (John Lilly's *The Center of the Cyclone*) gave a series of numerical "nicknames" as specifications of the spectrum of possible human levels of consciousness. The series went as follows: 6,144, 3,072, 1,536, 768, 384, 192, 96, 48, 24, 12, 6, and 3.

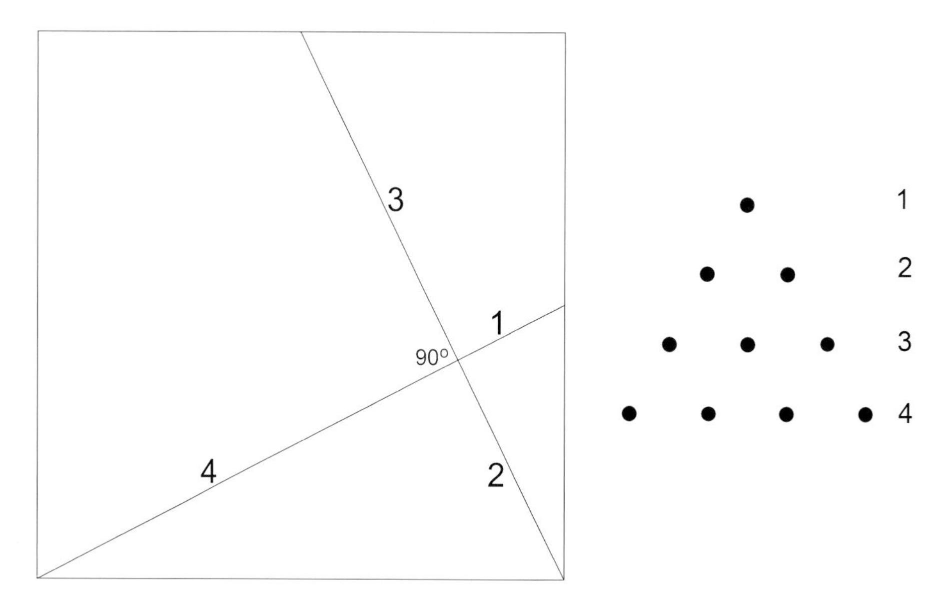

Figure 18.7

There is a substantive similarity between two mid-side diagonals of the Starcut and the Tetraktys, though the information is differently deployed.

The first was the most dense, the last the most translucent. I had first come across this number series applied, rather cryptically, by Gurdjieff and Ouspensky, to "hydrogens." Ichazo's information was more developed and included material that connected interestingly with work of my own, such as his reference to the number 24, in the sequence above, as the number of pure life. I took this to mean that while there is a body pulse of 72, there is a life pulse of 24.

To explain what I mean by this: any Fibonacci-type number series will be found to have a rhythmic pattern of 24 running through it. This is because if each number in the sequence is compressed (by adding the individual integers within it until it resolves to a single integer) the series reveals a repeating pattern; after 24 steps an identical set of compressed numbers appears. This will continue to infinity.

Now referring back to Critchlow's information about the lambda (which information is coded in musical terms on the board being held up for Pythagoras in Raphael's *Causarum Cognitio*), it was essentially an arrangement of numbers (figs. 18.8–18.11).

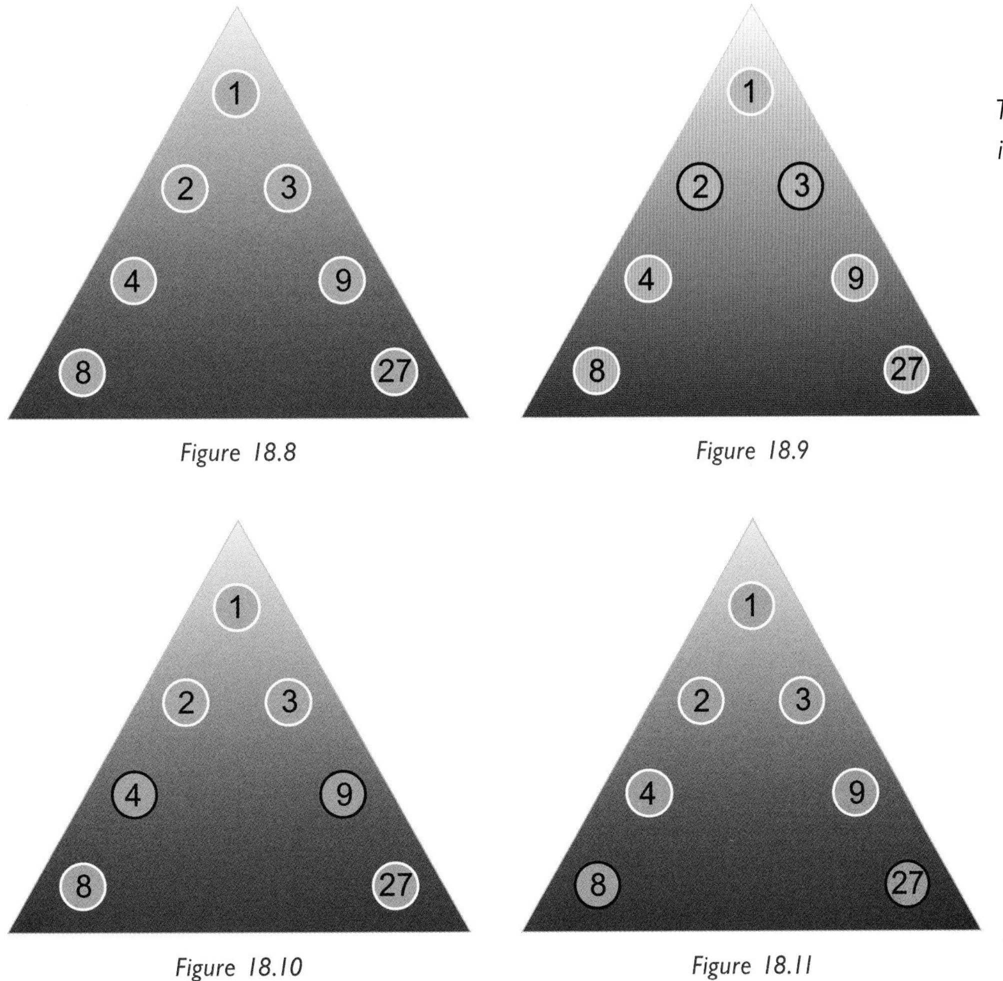

Figure 18.8

Figure 18.9

Figure 18.10

Figure 18.11

Figures 18.8–18.11. The numerical relationships in the lambda diagram.

One sees immediately that the numbers multiply out from 1; multiplying by 2 downward to the left and by 3 downward to the right (fig. 18.8). The numbers 1, 2, and 3 therefore determine the system (fig. 18.9).

Square numbers, 4 and 9, appear immediately below the three principles (fig. 18.10); and cubic numbers, 8 and 27 come below these (fig. 18.11).

To complete the Tetraktys form numerically, one needs to add the extra numbers 6, 12, and 18 (fig. 18.12).

Figure 18.12

All numbers that are times-2 multiples are in "octave" relationships with each other. Taken as musical frequencies on a string, they go up an octave when doubled and down an octave when halved. Notice the series 3, 6, 12 that, if extended by ten more steps will arrive at 6,144.

It will be seen that 6, 12, and 18 obey the multiplication system already implied. All multiplications downward to the left are by 2 (thus are octaves) and to the right are by 3. This simple table is a second order of number in the model—the first order being the pebbles themselves in their array: 1 + 2 + 3 + 4 = 10.

That first order establishes fundamentals such as the spatial potentials as in figure 18.13.

a point,
becoming a one-dimensional line,
making a two-dimensional plane (the simplest being a triangle),
extending to the third dimension with a fourth point that establishes a tetrahedron.

That first order of number comes from the addition of 1. The second order, suggested by the lambda, comes from multiplication by 2 and 3.

From my interest in Ichazo's work, I wondered what would happen if the lambda/Tetraktys was extended so as to see the whole table of numbers that it would generate if the second downward column to

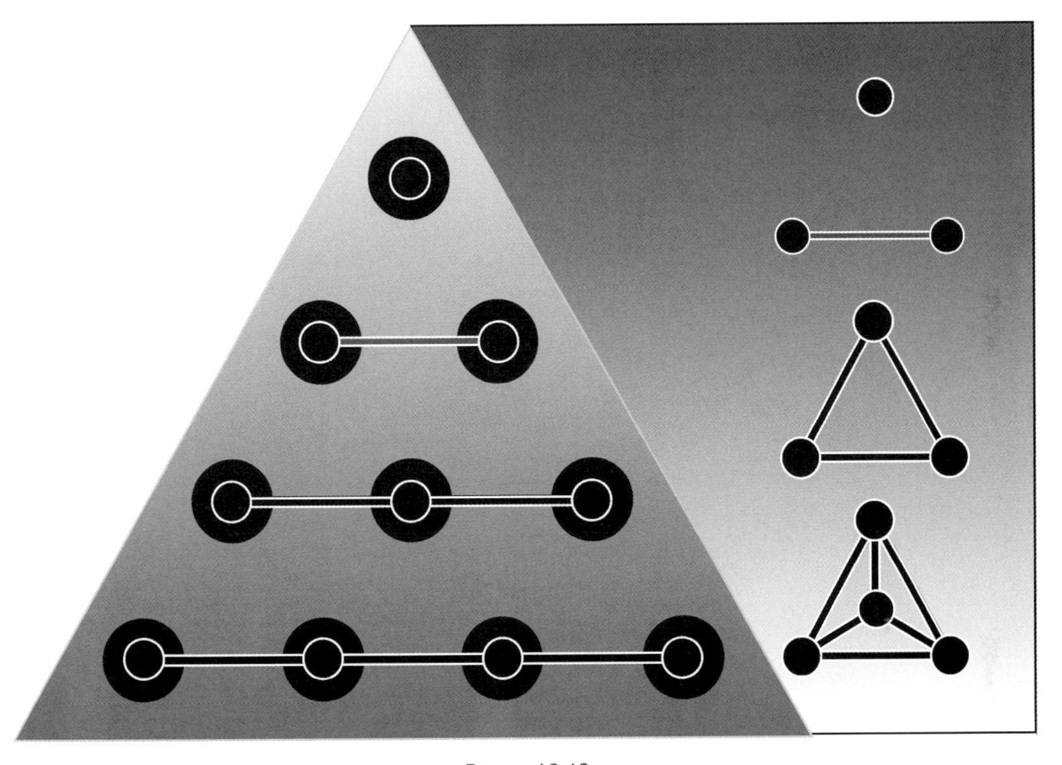

Figure 18.13

A point becomes a line that with a third point makes a two-dimensional plane that with a fourth point extends into the third dimension.

the left, with the series starting 3, 6, 12 was carried on right through to 6,144.

As with the Starcut, there was more than I had bargained for.

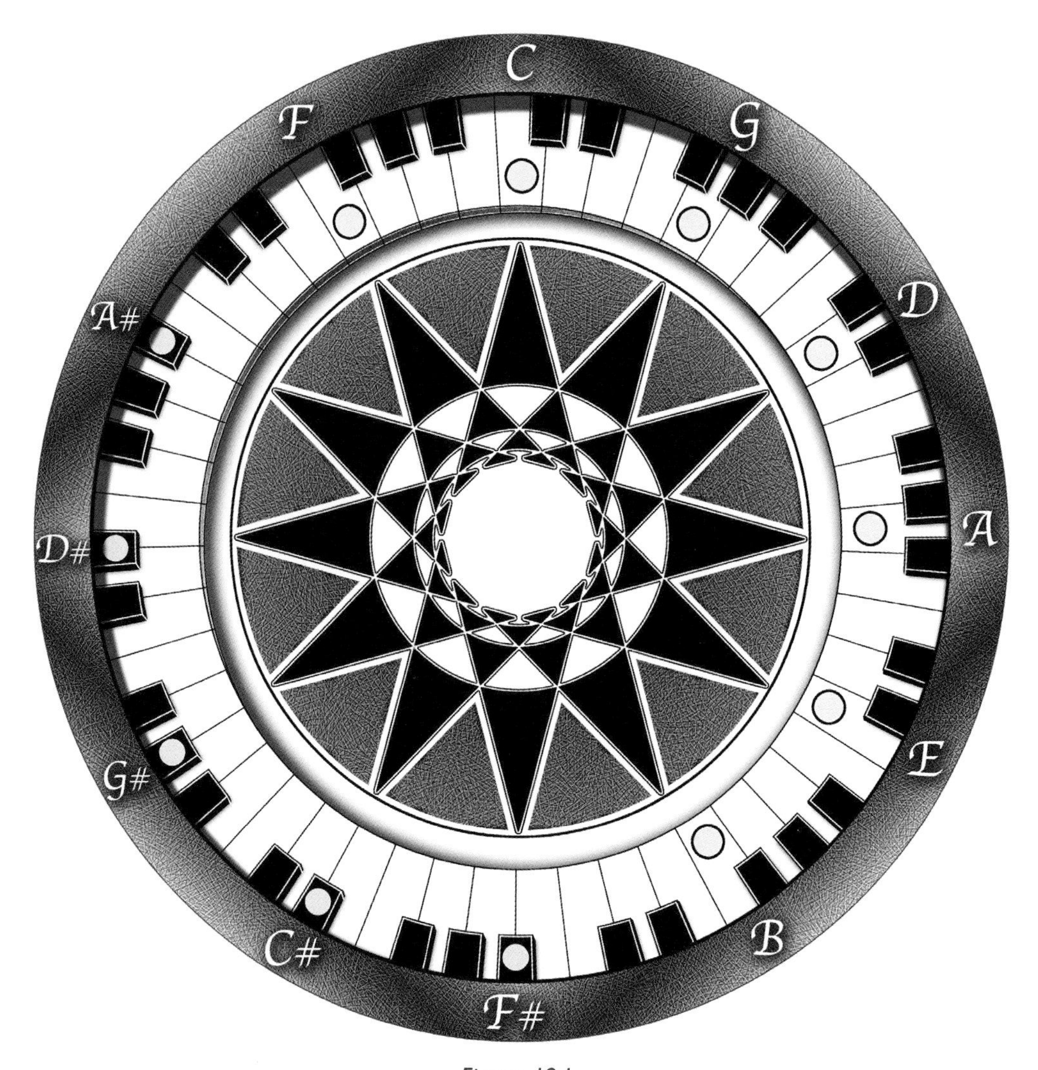

Figure 19.1

This circular piano keyboard extends for seven octaves, by which point the set of ringed notes comprises all twelve notes of the chromatic scale. Going clockwise around the circle, the ringed notes ascend in harmonic intervals known as perfect fifths. The interval sounds like the first two notes of the bugle call "The Last Post." The descending anticlockwise intervals are known as perfect fourths—for example, the first two notes of "Born Free." If one follows the line of the twelve-pointed star from C at the top, as it reflects back and forth within the circle, one gets either C, C♯, D, . . . , etc., or C, B, A♯, . . . , that is, one moves by semitones, either up or down (these notes occur, however, in different octaves). The "cycle of fifths" is said to have been the source of the twelve-tone chromatic scale since it lands on notes with twelve distinctly different pitches before arriving at a note that is the octave of any of the previous ones. The thirteenth interval brings it to a note that very closely approximates to that from which it first started (more on this below)—but now seven octaves higher. If one goes anticlockwise the journey descends by fourths rather than ascending by fifths.

19
Harmony at the Horizon of Being

From this Intelligible World, replete with omniform ideas, this Sensible World, according to Plato, perpetually flows, depending on its artificer Intellect, in the same manner as shadow on its forming substance.

Thomas Taylor, introduction to *The Philosophy and Writings of Plato*

The result of enumerating the Tetraktys to include the spectrum from 3 to 6,144 was daunting at first sight but turned out to have fascinating features, many of them shared by the Starcut square. The honeycomb effect in figure 19.2 (p. 198) helps in spotting certain relationships.

Some simple musical theory, already coded in the lambda, emerges more clearly. This "pyramid" reveals itself as being, among other things, a table of significant musical relationships. In the upper set of 10 (those with the palest background) one sees, as mentioned above, the octaves in the numbers 1, 2, 4, 8, ... This 1:2 ratio is in all left downward columns throughout the figure. Notice that the eighth move outward from 1—the unity—down the left arm of the figure is 256. It is an "octave of octaves" therefore. Interestingly 256 was for some time in

Europe the frequency for middle C, and was known as "philosophic C." Accordingly I have denoted the leftmost downward column as C. It defines that column's note.

When we compare the actual sounds that come from the singing or

Figure 19.2

The numbers can be taken as musical frequencies (or hertz, vibrations per second) so keynotes are given at either end of sloping columns identifying each as a set of octaves of the same note. Each column, through always being in 2:3 ratio to the column to its right, can be seen to be in ascending perfect fifths.

playing of any note, followed by a note an octave higher, we find that both are "the same note" tonally, even though we hear one as higher than the other. Octaves are mysteriously both the same and different. If a guitar string is stopped halfway along its length, it will sound an octave higher than it sounds if struck when open. Its vibratory frequency (or hertz: vibratory waves per second) will double as its length is halved. The relationship is essentially 1:2 and is the basis of all musical tuning.

Those knowledgeable about music will find what follows oversimplified. Mathematical tuning immediately brings us to the Pythagorean *comma*. This is a slight frequency discrepancy that occurs as one applies exact whole-number ratios to musical pitches. The sounds of the notes become increasingly dissonant the further apart they are. The comma was ironed out by what we call "equal temperament." This allowed the development of the harmonic/symphonic music of Western culture by making it possible to play chords including notes that are widely spaced from each other. Equal tempering is a compromise that, while troublesome to some highly refined musical ears, allows us Bach, Beethoven, and the Beatles. Later I will show that the pyramid of numbers used here facilitates calculation of the comma with mathematical accuracy, but for now here are some whole-number generalizations.

If you pluck a guitar string it will give the sound of the string vibrating "as one piece." Inside that obvious note, and usually unnoticed (but detectable by some people and by acoustic devices) will also be the sound of the string vibrating in two equal sections. That is the octave sounding, and it will be there because the nature of vibration is, quite spontaneously, to form waves within waves. The whole string is a long wave in which a half-wave also forms. It is secondary and quieter but it is there. A further sub-wave—quieter still—will be the sound coming from the threefold division of the string. If that sound is amplified it will be found to be a note that is not the same as the fundamental note of the whole string; but it is in harmony with it. On any instrument play the note C and the G above it, or hum the first two bugle notes of "The Last Post." The interval is 2:3. If you play an open guitar string then play the same string stopped at the seventh fret above it on the fingerboard, you hear this interval because one-third of the string

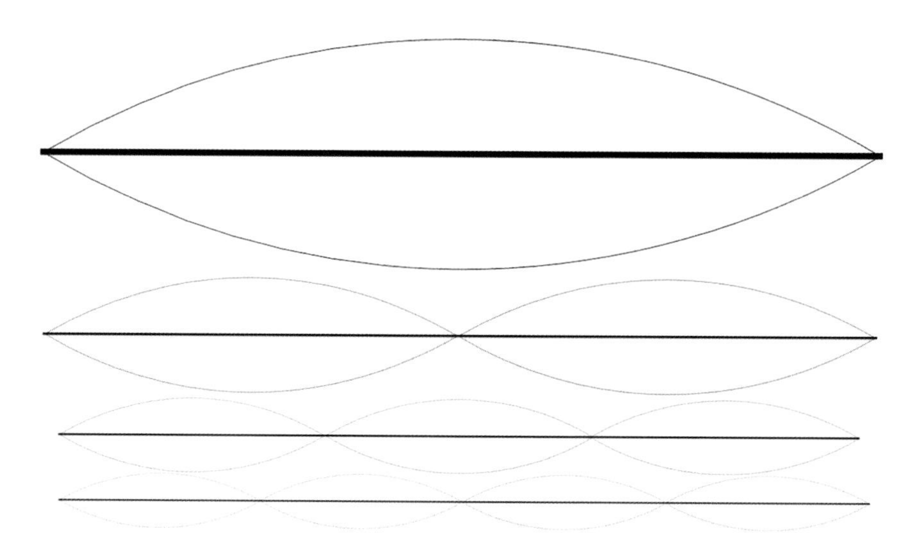

Figure 19.3

The top line is the whole string vibrating. The lower lines are the vibrations of various proportions of the whole string. Precision acoustic devices can detect these "partial" frequencies, and more, in the sound of any vibrating string. Musical theorists assume that in principle the wavelets of vibration continue, diminishingly, to infinity. The wavelets closest to what the ear can hear, when played together, sound harmonious to us.

has been silenced. Two-thirds of the string are now sounding. This is known as a "perfect fifth," 2:3. C to G is an example. Played together, they harmonize (fig. 19.3).

Looking again at figure 19.2 and at the diagonal line of letters going down from the top to the right (C, G, D, A, E, etc.), and for now ignoring all the numbers, one sees that C and G start a process that goes through the sequence that was given by the ringed notes in figure 19.1. They denote the cycle of fifths.

The "fifth" ratio 2:3 is in the second row of the numbered Tetraktys. The arrangement of the Tetraktys is such that *any* two numbers adjacent to each other, in whatever row, will always be in a perfect fifth relationship. So, for instance, 72:108 or 144:216 are in this ratio. There is a further standard musical harmony also shown here: the perfect fourth. If you stop a string so as to sound three quarters of it, it will play a perfect fourth above the sound of the open string. Play, for instance,

Figure 19.4
A few examples of perfect fourths are marked here in red.

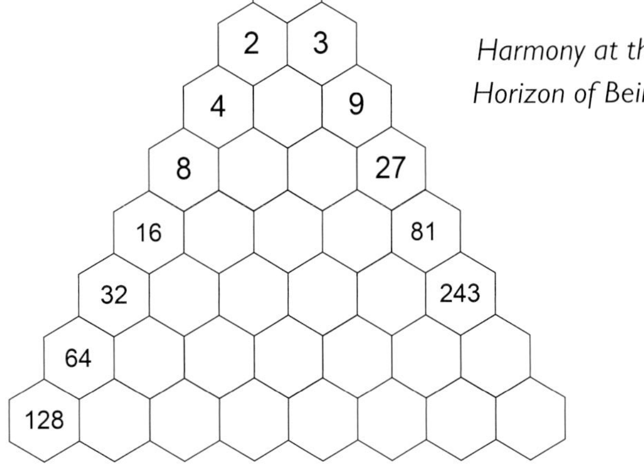

Figure 19.5
The numbers that are the basis of the classic Pythagorean tuning.

C followed by F. So anywhere within the numbered Tetraktys where two numbers (frequencies) are in the same relationship as are the 3 and the 4 in the second and third rows, you will find a perfect fourth. The interval between 108 and 144 is thus a perfect fourth.

In Figure 19.5 are included the numbers that are the basis of the classic Pythagorean tuning that was built up to a full octave scale through a series of fractions: 1, $9/8$, $81/64$, $4/3$, $3/2$, $27/16$, $243/128$, 2.

The numbers in the Tetraktys, many of which are the same as those in the Starcut, continually remind one of the "cosmograms" and number lore of different traditions. Keith Critchlow has pointed out how the three median relationships ("mean ratios") of the Platonic tradition can be found within the first four rows of the numbered Tetraktys (figs. 19.6–19.8, p. 202).

What these mean points denoted were necessary implicit points of mediation between a principle and its manifestation in the form of its octave—the up or down scaled reflection of itself.

If one thinks of the electromagnetic spectrum, in the moments of its coming-into-being as the total field of vibration, one may ask "What gives it its form? How come anything fixed is generated? How come

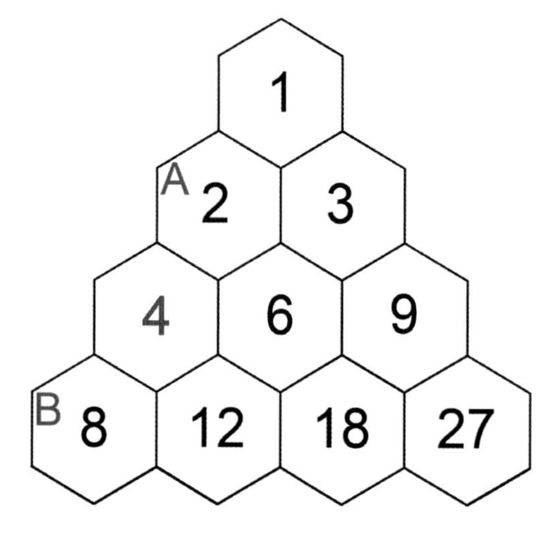

Figure 19.6
The number 4 is the octave between 2 and 8, and is their "geometric mean"; mathematically it is $\sqrt{AB}$.

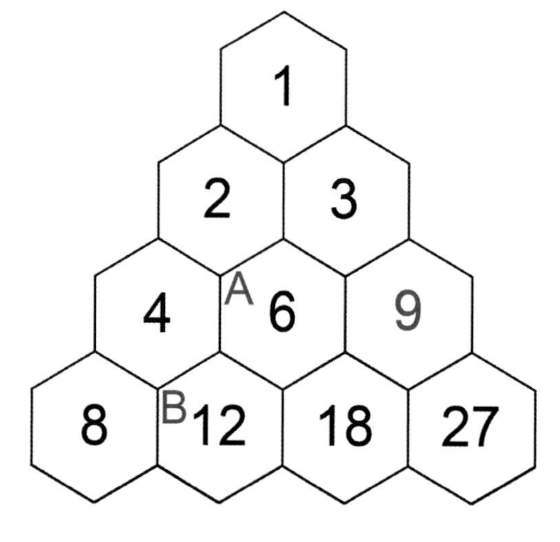

Figure 19.7
The number 9 is the perfect fifth between 6 and 12, and is their "arithmetic mean"; mathematically it is $(A + B)/2$.

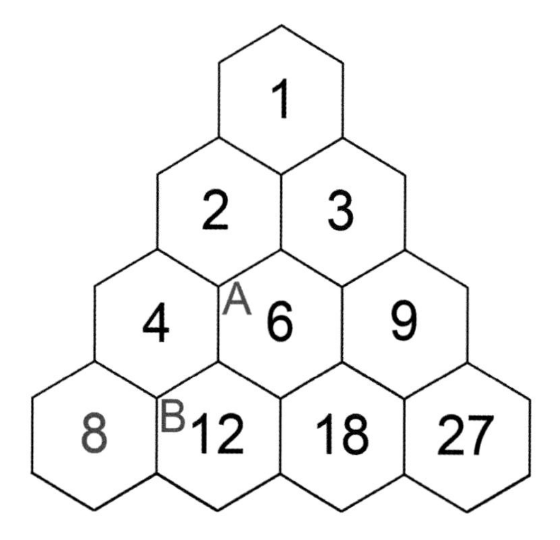

Figure 19.8
The number 8 is the perfect fourth between 6 and 12, and is their "harmonic mean"; mathematically it is $(2AB)/A + B$.

there are points of manifestation where effects embody their causes at a different level?" For the Pythagorean—and traditions such as that of the *Sepher Yetzirah*—number itself was considered the operator, and the conjugal relationships between numbers were the rationale of all manifestation. As David R. Fideler writes in his introduction to *The Pythagorean Sourcebook*:

> The solution lies in the limiting power of number . . . the problem is one of mediation or harmonia through the medium of numerical proportion.

In an article "Musical Theory and Ancient Cosmology" posted on the internet in 2003, musicologist Ernest McLain refers the roots of this harmonia right back into the mythology of ancient Mesopotamia:

> Theology, from its birth as "rational discourse about the gods" and in many later cultures influenced by Sumer, is mathematical allegory with a deeply musical logic. Tuning theory today remains a fossil science with no change at all in its basic parameters, structured by the gods themselves in numerical guise, since it premiered in Sumer about 3300 BC.

All the relationships illustrated in the graphics above will be found throughout the whole extended numerical Tetraktys. All numbers in equivalent juxtapositions will be found to be related in equivalent ways. One can often use the extended Tetraktys to read off musical proportions that appear in the measures of temples, monuments, and symbolic structures. The liberal arts traditionally included both music and architecture under the one heading: Harmony. From this perspective one can understand why.

THE CENTRALITY OF 6

We have already seen a lot about the sexagesimal system appearing in the Starcut with its two layers of number, the first of which results in

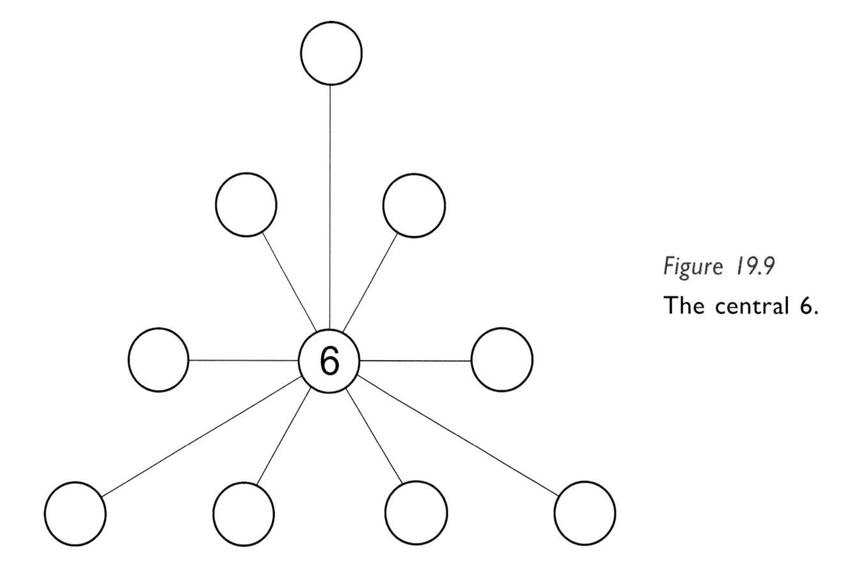

Figure 19.9
The central 6.

the big triangle in the square as having a value of 6, and the second, more detailed, number system giving it a value of 216. The numerical Tetraktys shares these sexagesimal characteristics.

It is not until the number 6 appears in the middle of the third row in the numerical Tetraktys that the rationale of the numbering system is made plain. The prior sequence goes 1, 2, 3, and 4 and could simply mean that the figure is to be numbered sequentially from 1 to 10. In that case the position of the 6 would be occupied by the number 5. The fact that it is a 6 means that one has not in fact been adding at all. The diagram is essentially to do with multiplication.

Very early on in this book we came across fourfold-ness as representing the notion of a manifest result. In the primary vesica piscis it is a rhombus that picks up all the points generated by the circles. Here it is a rhombus of number, rather than the seed triad, that reveals the multiplication process. The number 6 is further appropriate because it is the perfect number of circularity, owing to the sixfold division of a circle's circumference by its radius. If one does some pattern recognition on these lambda numbers in the Tetraktys, 6 becomes predominant. All triangles of adjacent numbers with their point upward are divisible by 6. Of course the seed triangle of 1, 2, and 3 is the root of this. It has beautiful results. Here are a few.

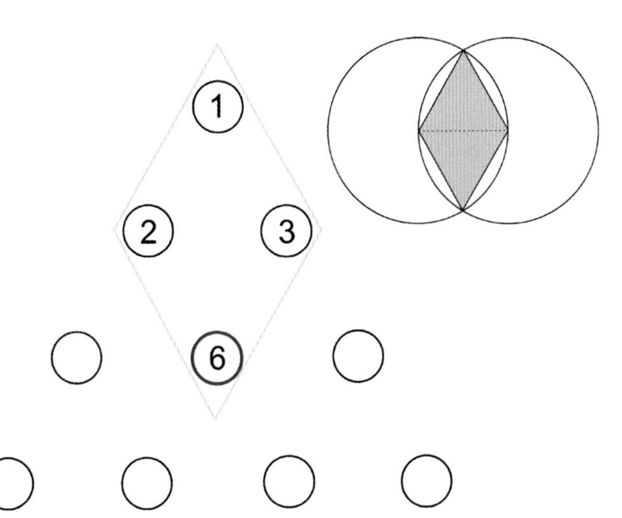

Figure 19.10
It is the 6 that shows that the multiples 2 and 3 are at work. Geometrically 1, 2, 3, and 6 stand in a rhombus relationship. They are also, of course, the omni-perfect number sequence, as we have seen previously.

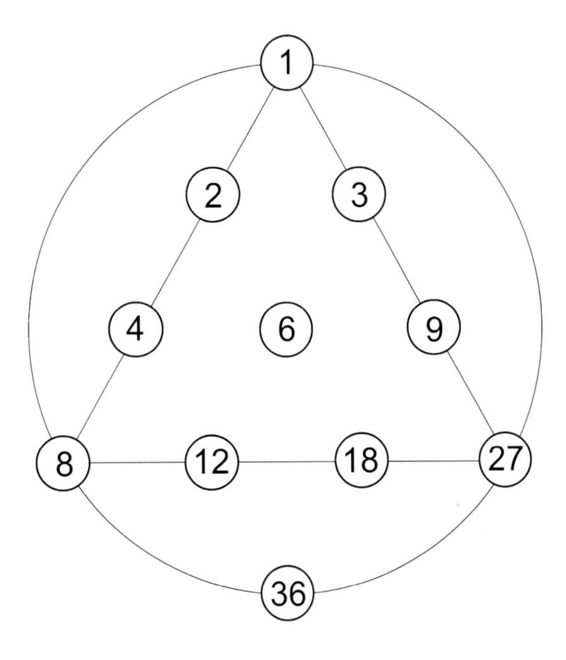

Figure 19.11
If the triangle of the primary Tetraktys is encircled, one further number from the extended Tetraktys falls on the circumference: 36, or 6^2.

2	9	12
36	6	1
3	4	18

Figure 19.12
Magic squares normally reveal their magic by processes of addition. This one does it through multiplication. All eight lines crisscrossing the square multiply to 216. Of the eleven numbers in the previous figure (19.11), 8 and 27 are not included in the square. We will see shortly how they reenter the game.

If the triangle of the primary Tetraktys is encircled as seen in figure 19.11, one further number from the extended Tetraktys falls on the circumference: 36, or 6^2.

If we apply a "magic" multiplication square of 216 (yes, inevitably there is one!) to the triangular geometry of the Tetraktys we find harmonies in plenty. First consider the square.

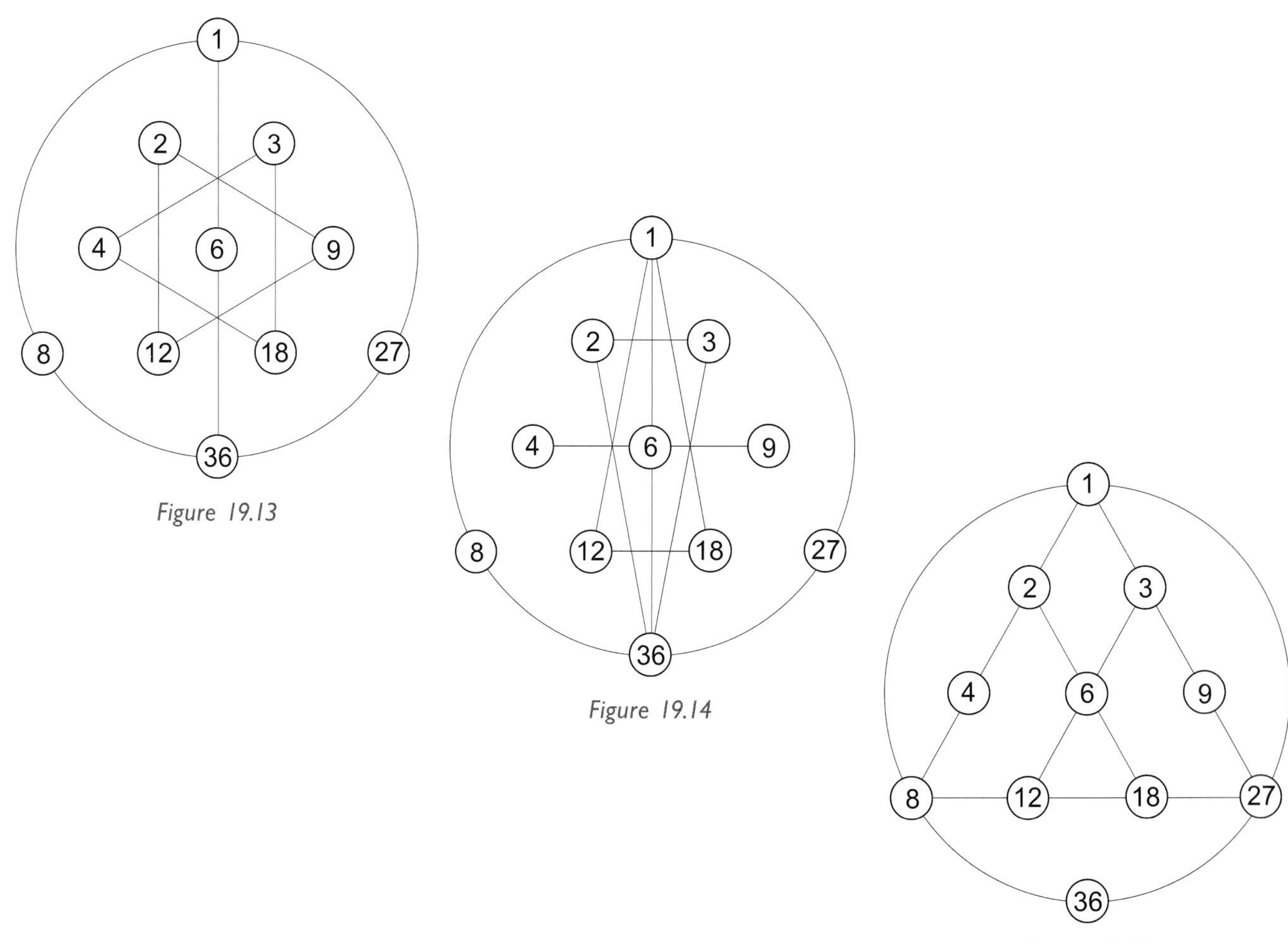

Figures 19.13–19.15. The products of the numbers connected as triangles in figures 19.13–19.14 are uniformly 216. Triangles with this same product could also be drawn between 8, 3, and 9, and 27, 2, and 4. The vertical lines have the same product; as do the diagonally connected numbers in figure 19.15. Also the 8 and 27 that are missing from the magic square combine with the 1 in a boundary triangle that produces the same total: 1 × 8 × 27 = 216.

We find a correspondence between the eight lines of the square and the number patterns that occur in the extended numerical Tetraktys, which turns out to be swarming with 216.

In view of this it comes as no surprise to find that the extended Tetraktys' centermost number, in the middle of the seventh of its thirteen rows, is also the ubiquitous 216 (see fig. 19.2, p. 198). Its musical note, incidentally, according to the logic established by 256 as a "philosophic C," is the note A, one octave below the "standard orchestral pitch for A" that used to be 432 hertz but was changed early in the twentieth century to its present American orchestral standard of 440 hertz.

Some further musical information that can be derived from the big numerical Tetraktys is the exact Pythagorean comma—that small discrepancy, previously mentioned, between the outcome of a strictly mathematical cycle of fifths, based on ratio 3:2, and a seven-octave cycle based on a ratio of 2:1. If one divides by 2 the bottommost right-hand number 531,441 in figure 19.2 (the reappearing note C), after nineteen such divisions the result is 1.0136433 instead of 1. The numbers after the decimal are the "comma" itself. It is that slight discrepancy that causes dissonance in chords that include notes that are some distance from each other. The comma was averaged out across all the notes by the "equal temperament" tuning method that is now standard in European music, having been famously adopted by J. S. Bach and used across all twelve keynotes in his *Well-Tempered Clavier.*

PRIME NUMBERS

The numerical Tetraktys contains only the primes 2 and 3. The number 1 is, in technical mathematical parlance, not a prime—indeed not a number at all—but is the basic unit from which numbers are composed. But by adding groups of numbers into linked sets (examples are shown in figs. 19.16–19.27) one gets, I conjecture, *all* primes and their multiples. "From 1, 2, and 3," say the Pythagoreans, "come all numbers." Correct.

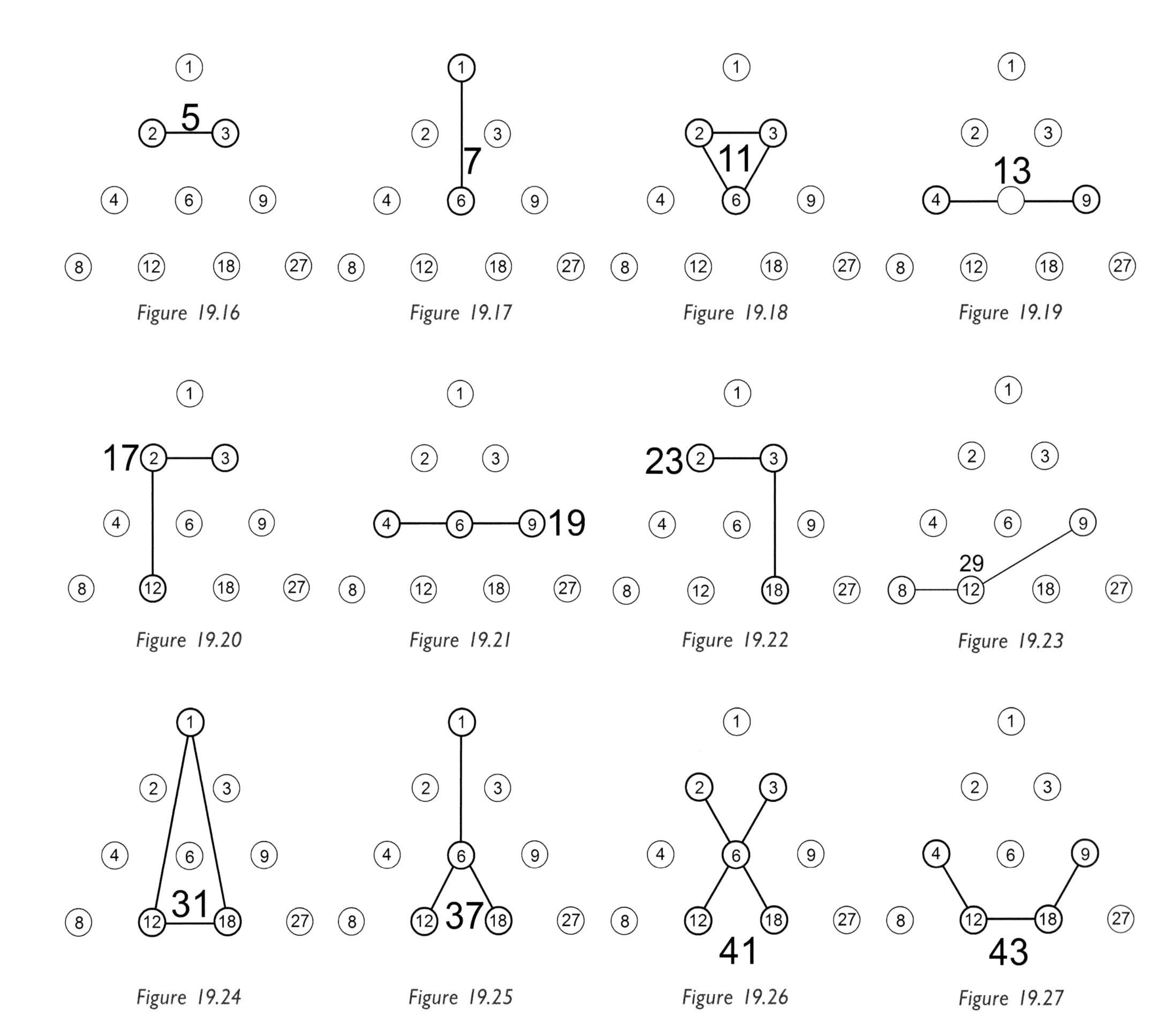

Figure 19.16 *Figure 19.17* *Figure 19.18* *Figure 19.19*

Figure 19.20 *Figure 19.21* *Figure 19.22* *Figure 19.23*

Figure 19.24 *Figure 19.25* *Figure 19.26* *Figure 19.27*

Figures 19.16–19.27. Linked sets of values in the Tetraktys. The first 5-set is the 2 and the 3 on the second row. All pairs of adjacent numbers throughout add to a multiple of 5, and the number between and above them tells you which multiple of 5 they are (fig. 19.16). Two numbers vertically adjacent will come to a multiple of 7 and will indicate, by the upper number, which multiple it is (fig. 19.17). Where there is a blank ring, the number in that position is excluded from the sum (as in fig. 19.19). So, for instance 12 + 27 (with the 18 excluded) is the third multiple of 13.

The little "set-shapes" are movable anywhere, always denoting their characteristic prime. Multiples of primes will have the same shape. One can check which multiple of a prime by seeing how the number 1 stands in relationship to the original shape. In the case of a multiple, another number will be in that relative position. That number is the multiple. See figure 19.28 on the following page for a couple of examples.

My conjecture is that there can be found shapes for all primes in some endlessly extended numerical Tetraktys, and various different shapes can result in the same prime.

Unfortunately all this does not promise a solution to the search for a predictive theory of prime numbers because trying to find what shape might come next is just as difficult as predicting what prime number might.

Having seen how the primes occur as multiples, recognizable in similarly shaped linked patterns, we realize that this applies equally to those patterns of 216 identified earlier; and that this entire mountain is seething with that number.

There is much else that can be found in the extended numerical Tetraktys. There are clusters of numbers such as one including all the numbers that occur in the degrees of angle in pentagonal geometry: 36°—the angle at the points of a star pentagram; 72°—the angle of the star's isosceles triangles; 108°—the angle between the sides of a pentagon.

There are further glosses that can be made concerning this numerical device. The numbers ascribed to Peter (486) and Cephas (729) in the Greek gematria code that was detected in the New Testament by Frederick Bligh Bond, will be found adjacent to each other. A cube 9 × 9 × 9 has a facial surface of 486 and a volume of 729. This was quoted by Bligh Bond as an example of a numerical pun in the New Testament, visible only to the initiated.

In terms of the Tetraktys this could be just an obscure coincidence. However, the number 729 occurs, contrasted with the number 1, in an allusion that Plato makes to the contrast of the "Good Man and the Tyrant." Within the numerical Tetraktys these numbers do take on a significant relationship of opposition. If 1 is the note C, then 729

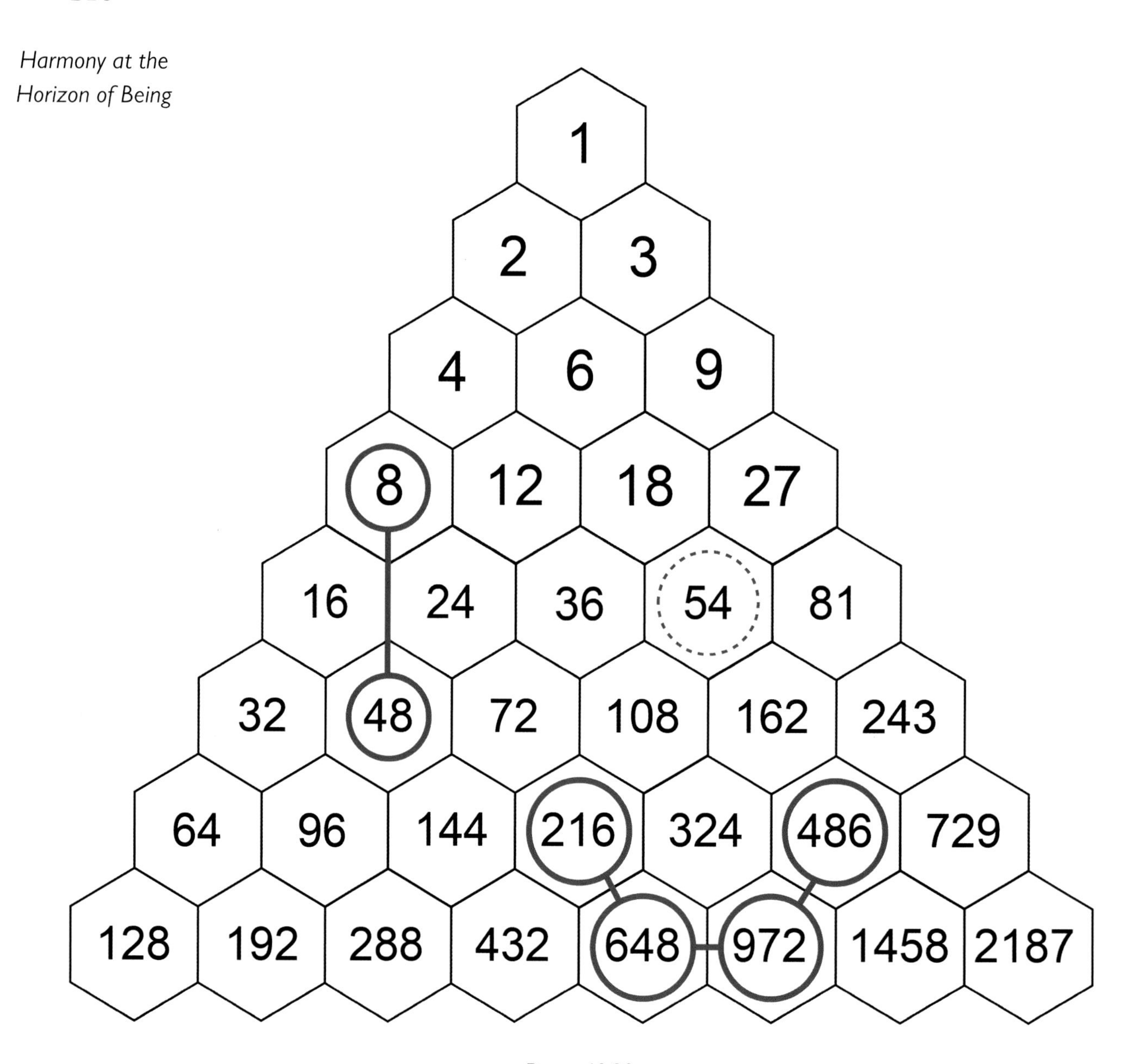

Figure 19.28

The set with numbers 216, 648, 972, and 486 adds to 2,322, which is 43 (the shape in fig. 19.27) × 54—which stands in the relationship to this set as 1 stood in relationship to the first occurrence of 43.

(as can be seen in the extended pyramid of tones) is the note F♯; eight octaves distant. These two notes are in what is known as a "flattened fifth" relationship to each other—a famously inharmonious interval—and further distorted by the discrepancy caused by the Pythagorean comma. This relationship of the flattened fifth was thought to be so dissonant that it was, for a long time, forbidden in Church music, and was known unflatteringly as the *diabolus in musica*—the devil in music.

All in all, when contemplating the Tetraktys in its extended lambda form, as when contemplating the Starcut and its numerical aspects, I find it hard to believe that others have not passed this way before. Whether they have done so or not, I find it strangely moving that, completely unsought by myself, there should be this same mysterious—and seemingly long forgotten—number of 216 at the heart of the numerical Tetraktys, at the heart of the Starcut itself, also as the measure of Pythagoras's perfect cube, as the Kabbalistic number of the oracle of the Jewish Temple, and as the value of the primary trigram of Chinese philosophy.

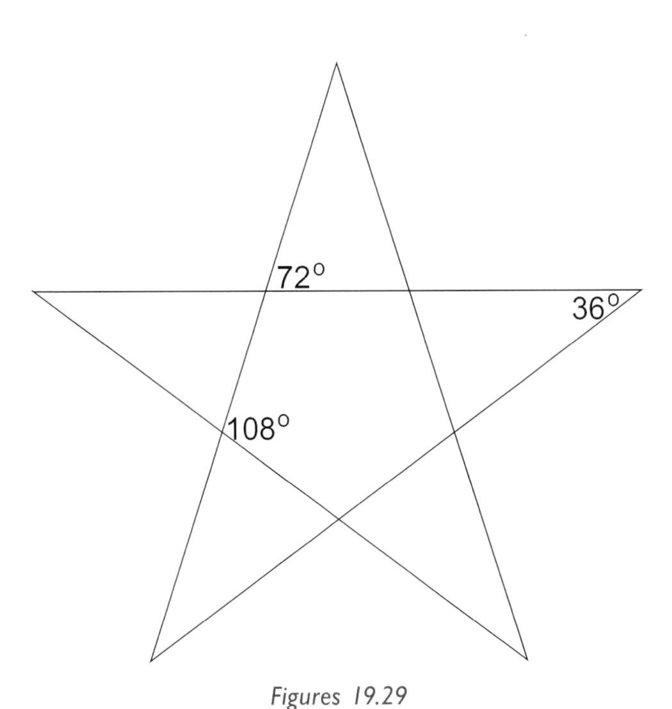

Figures 19.29

Angles in the geometry of the pentagonal star.

Figure 19.30

A scientific model of the DNA spiral, looking down through the structure. At the points of the ten-sided regular decagon the reflex angle is 216°—thus the figure's outer form is contained within a total of 2,160°. One recalls the words from the Pythagorean vow: ". . . holy Ten, sacred key holder of all."

Figure 19.31

Fields of number: A sphere, drawn using threefold geometry, on a patchwork derived from the Starcut.

20
The Shape of Concord

In the Middle East among Orthodox Christians and Armenians, there is a custom that they do not use an organ in Church; they use a chord or sound made by ten or twelve persons sitting with closed lips. Any one who has heard it will say that they are right; the sound of the organ is most artificial in comparison. The sound produced by the voices of ten or twelve persons with closed lips . . . reaches so deep and far into the heart of man . . . it is a natural organ which God has made.

Hazrat Inayat Khan

In the late 1950s, my mother, tired of hearing my strident renderings of "Basin Street Blues" on the trumpet, bought me a guitar, and from that I learned quite a bit about harmony. Accompanying songs tends toward thinking in chords—sets of notes played together. There are also sets of chords that tend to cluster and commonly occur in the same song; and standard left-hand fingerboard technique also emphasizes basic positions and movements. Overall, certain patterns become very familiar. All this was long before I had any particular interest in geometry.

Opposite:
Figure 20.1
The Concert,
Vermeer, 1664.

Years later, studying the Starcut, I wondered what might happen if the number set 3, 4, 5, that adds to 12, was used as a way of sectioning the twelvefold chromatic scale of music. The result was a beautiful concord between number, shape, and harmony.

The Starcut diagram gives a convenient way of drawing a regular twelve-sided polygon (a dodecagon). It is not perfectly accurate; but since it has an error of only one two-hundredth of a degree on eight of its points, it will do quite nicely.

The perimeter of a 3-4-5 triangle measures 12 as does the circular face of the chromatic scale of music, I looked to see if there was some correspondence. The results speak for themselves. A geometric rationale for the entire range of harmonic triads appears (see fig. 20.3).

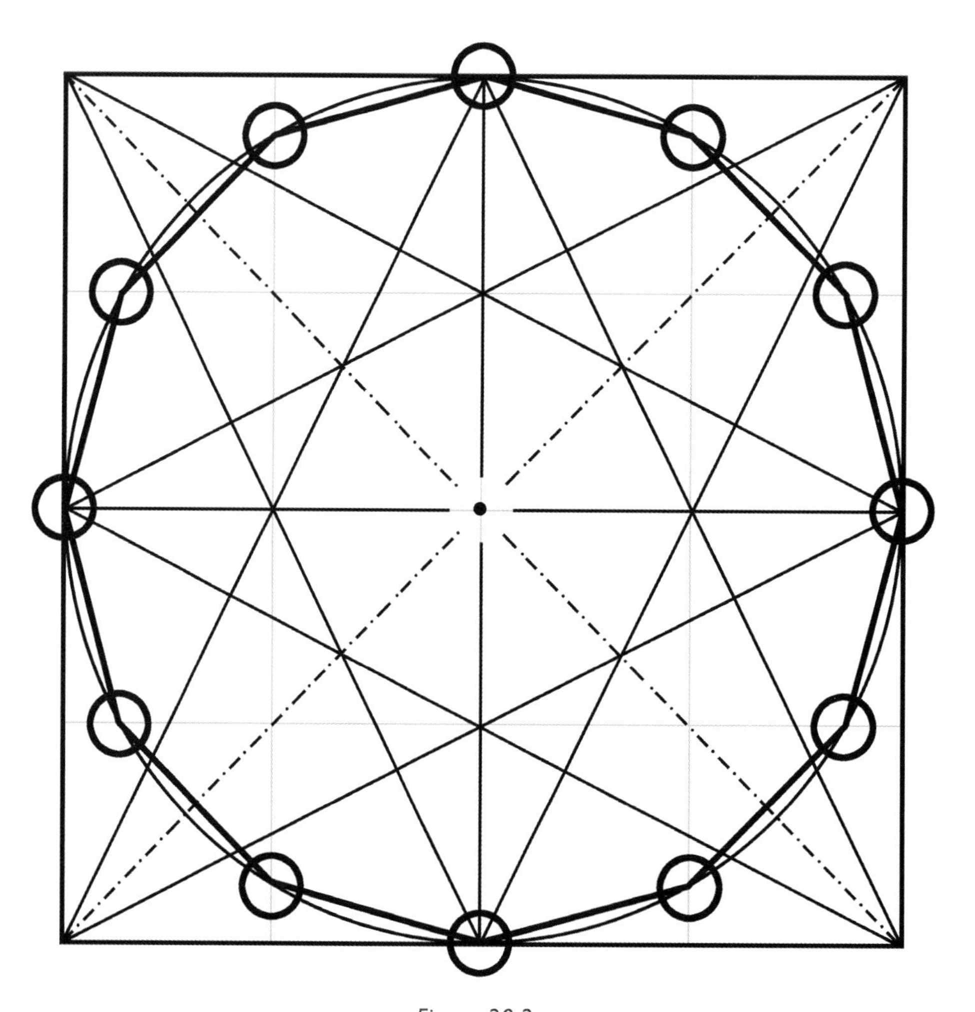

Figure 20.2
The dodecagon.

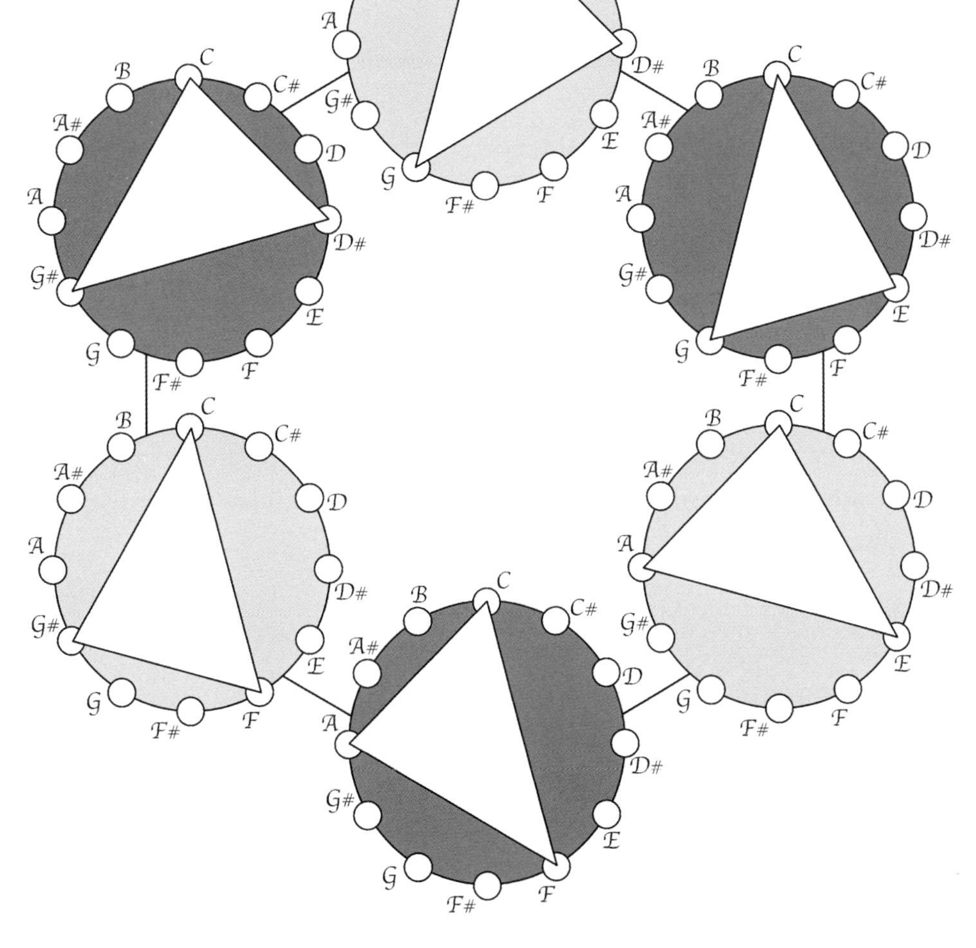

Figure 20.3

The twelve-note chromatic scale is repeated six times here, allowing all the combinations of the 3, 4, 5 number set to be used. Throughout, the note C remains constant as one of the vertices. If each triangle of notes is played as a chord, all the major and minor triads containing the note C are here. The clockwise progression used allows for just one note changing position at each step as one goes around the hexagon. The three minor triads (paler background) and three major triads interlock just as do the two triangles of a six-pointed star. In segmenting the circles the 3-4-5 quality of the triangles has flipped from being the triangle's side lengths to being its angles, which are now the *set* 3, 4, 5: 45° (3 × 15), 60° (4 × 15), and 75° (5 × 15). Harmony of the spheres . . .

As a onetime folksinger, I was more than familiar with the relationships of the actual sounds of the chords that came from selecting notes according to combinations of the 3, 4, 5 set. This was the basic stuff of triadic harmony. A detailed analysis would show that the frequencies involved in the pitches of the notes all map beautifully on to the numerical Tetraktys (using old-style middle C at 256).

The hexagon of chord circles in figure 20.3 has only the six chords that include the note C as a tone in the chord. So what about the other eleven notes of the chromatic scale? It was in looking for them that I found the structure shown in figure 20.4.

Essentially there are just twelve hexagons in the array. The fact that they begin to repeat both up and down and from side to side means that the twelve-hexagon lattice can be seen as wrapping around a torus or donut shape. (Thanks to my friend Elliott Manley for this last observation.)

The correspondence between musical harmonies and geometric forms made me curious about what a simple equilateral triangle, or a square, would make in terms of musical chords. Results are shown in figures 20.5 and 20.6. I have left out the note names since, as we now see, these geometric structures show interval relationships that can be applied to any keynote. The names "third" and "sixth" similarly apply to the relationship of a note to its keynote (or tonic); as did the terms "fifth" and "fourth" that were used earlier. Musical nomenclature is quite a tangle. It sounds better than it reads!

The relationship between musical harmony and geometry suggests all sorts of creative possibilities. Previous publication of parts of my work on this has been the occasion of musicians getting in touch about results they have had through letting the geometry inspire musical ideas.

The relationship could work the other way too; for instance, the toroidal lattice within which geometric/musical chord shapes can be mapped suggests a three-dimensional environment of 144 lights programmed to the sequence of melody and harmony in any piece of music.

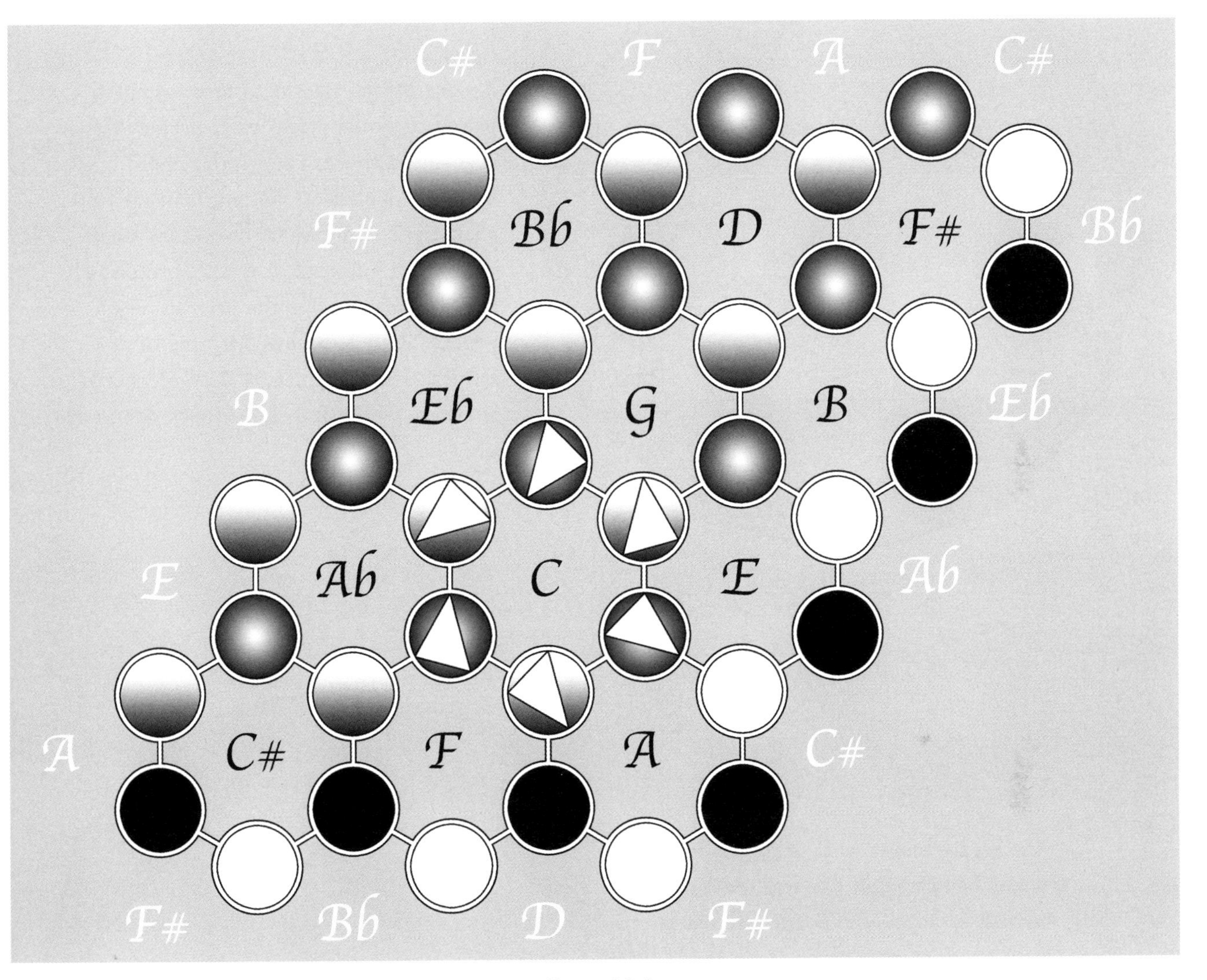

Figure 20.4

Allan Brown, who contributed this illustration, calls it the "harmonic honeycomb." The spheres lit from above are major triad chords, the centrally lit spheres are minor triad chords. All possible Western keynotes are present. The keynotes repeat (see the letters in white). The plain white-and-black circles denote that repeat process happening. It can be thought of as going to infinity, producing the same relationships.

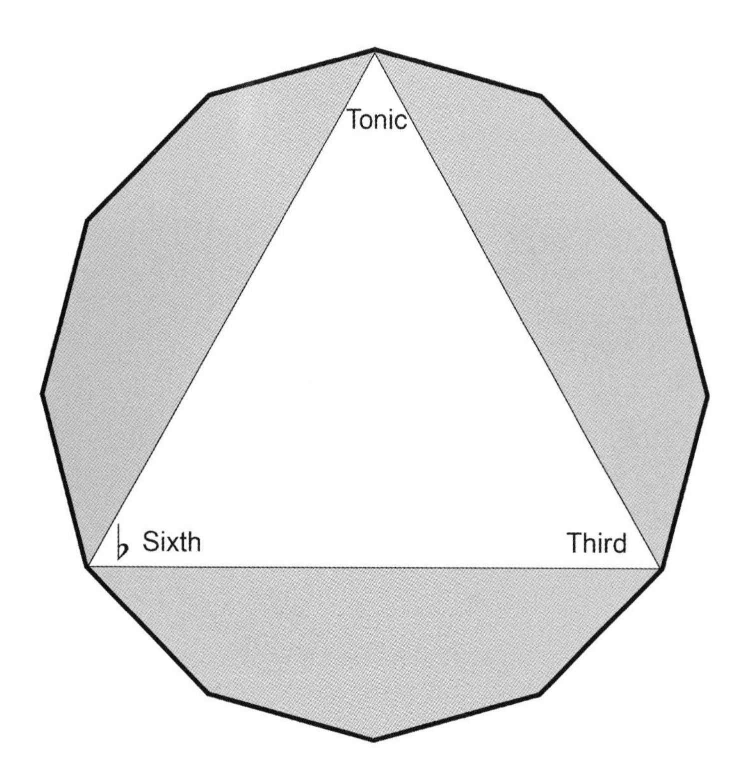

Figure 20.5

An equilateral triangle, as a set of notes played together, makes what is known as an augmented chord. Though this chord is to be heard in classical music it is much more characteristic of twentieth-century music with a jazz influence. It is an "unresolved" chord—meaning that the ear naturally expects it to be followed by one of the musical triads (the set of "3, 4, 5" chords shown in fig. 20.3, p. 217). Tunes don't end on it.

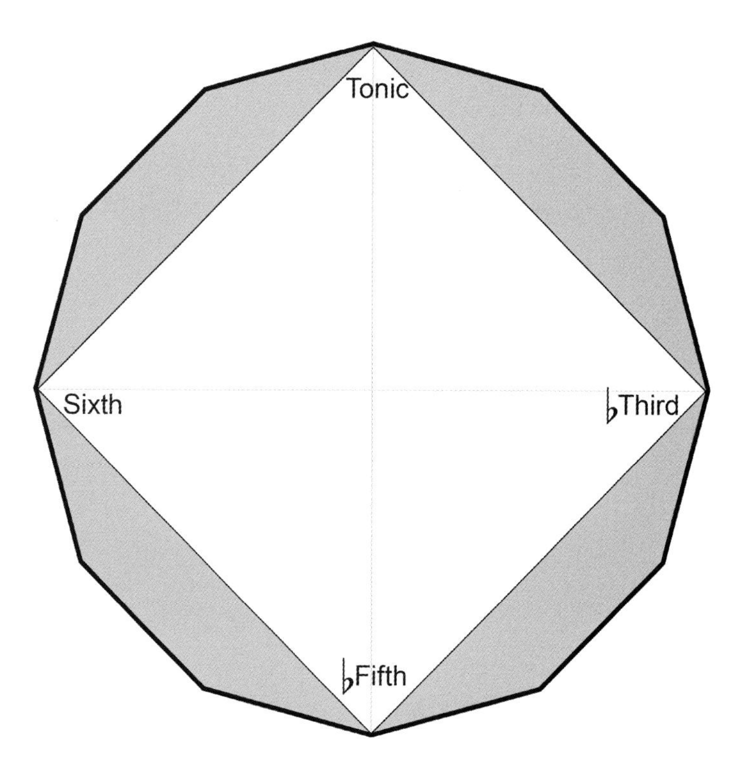

Figure 20.6

The square makes what is known as a diminished chord. It has two "devil's harmonies" in it—the notes that are immediately opposite each other are in this relationship. Any three of these four notes (making an isosceles right-angled triangle therefore) will sound as what is known as a "diminished seventh" chord. Both types of diminished chord are common in all music but, like the augmented chord, are unresolved.

Since the other three arts of the quadrivium—geometry, number, and music—are here working together it seems that their companion art, astrology, may have an entire harmonic aspect still to explore.

This is particularly so in terms of the 3, 4, 5 set. I am not an astrologer and have only a layman's knowledge of it. There seems, however, to be a glaring relevance to astrology in the foregoing. "Trines" and "squares" are considered to be significant aspects in horoscopes. It is said that they account for key dynamics in the person's life process. It seems to me that a birth chart might also be studied in terms of any 3, 4, 5 relationships that it happens to contain. Given that astrology has psychological meaning, such relationships should indicate highly stable aspects in a person's chart because they are musically "resolved"; just as squares and trines, being unresolved, indicate dynamic factors. Indeed any 3, 4, 5 aspects, logically speaking, should turn out to be the most significant constants in an entire chart. There should even be characteristics of general personal disposition that resonate with whether the triads in question are major or minor chords since these have notably different qualities of musical mood.

Figure 21.1

Between the towns of Draguinan and Brignoles in Provence, in southeastern France, stands the Cistercian monastery of Le Thoronet, famous for its beautiful acoustics, built in the late twelfth and early thirteenth centuries. It is said that the monks had to sing especially slowly and precisely because of how the architecture reflected their voices.

21
The Lyre of Apollo

The Starcut diagram swarms with musical information. As with the Tetraktys and the chromatic circle, musical proportions pervade both the geometry and the numeration. If one were to list the musical ratios between (for instance) the areas of different triangles in the diagram, the inventory would include all the triangles in conjunction with all the others. So what I will do here is highlight the musical aspects that are, perhaps, the most interesting—not least being the fact that this figure, which suggested the shape of a template for the Great Pyramid, equally directly suggests the structure of a musical instrument.

Further general points about geometry and sound will provide a context.

On a visit to Delphi, I stood in the center of the orchestra and spoke in a quiet undertone to Nora my wife, who was sitting high up in the bowl of seats. She could hear my words with perfect clarity. The acoustic qualities of Greek theatre geometry are famous. I have no doubt that the use of sound and sound effects goes far back into experiential processes involved with initiatic education. The human race has such an ancient association with caves and thus, presumably, the quite extraordinary acoustic effects that are to be heard in them. There are caves where a person a long way away may seem to be whispering in one's ear, or where, in some chambers one can shout and one's voice is simply swallowed up becoming inaudible even for people quite close by.

The "bowl" shape of the Greeks has been carried forward architecturally in a number of ways, and the aspect that I wish to discuss here is the architecture that characterized the twelfth-century Cistercian monastic movement. Having, many years ago, spent some time in a Benedictine monastery (the Cistercians are strict-order Benedictines) and, as a seminarian and priest, having spent many hundreds of hours reciting the psalms, antiphons, and prayers of the Divine Office, I have come to appreciate the intensity of the spiritual practice of the Cistercian discipline.

Bernard of Clairvaux, a seminal figure in the Cistercian order, was so famous as a preacher that when he came to town people were warned "lock up your sons." He was universally revered in his own time. His words lured men into crusades, into Templardom—which he promoted in its first ideals—or into the monastic life under the strict reformed Rule of Saint Benedict. He even successfully promoted the cause of one

Figure 21.2

The theatre at Delphi, Mount Parnassus, Greece. *Left:* immediately beyond the stage area (called the "orchestra"), down the hill, one sees a flat rectangle of ruins that are the remains of the temple of Apollo where the Pythia sat to give the oracle. The interpretation of her words was presumably announced to the parties in attendance via the ideal acoustics of the theatre arena. *Right:* a view from the orchestra.

of two contenders for the papacy. Bernard was clearly a man who chose his words; and there are two sayings of his that are, literally, resonant concerning the present topic:

> God is length, breadth, height, and thickness;

and

> No decoration, only proportion.

That he was true to his word is shown by the architecture he approved.

Nowadays, in almost all churches everyone kneels, sits, or stands in ranks of pews facing the sanctuary and altar. In this modern photo of the chapel (fig. 21.3), this is the way that the benches are arranged. However, the ancient conventual layout, still to be found in the "choir stalls" of great churches and cathedrals, had the participants facing each other. The following points about the build-up of sound and harmonics are accentuated when the singers are facing each other beneath a reflecting semicircular vault. Greek theatre acoustics are further intensified.

Figure 21.3

The chapel interior of Le Thoronet Cistercian abbey in Provence.

The reason that Bernard did not approve decoration but only the measures of appropriate geometry was (in addition to the reduction of visual distraction) because that way the sonic chamber was kept pure. The entire life-discipline of the Cistercians was structured around the singing of what is known as the Divine Office. In addition to the daily ritual of the Mass, the "hours" or sessions during any day would comprise Matins, Lauds, Prime, Terce, Sext, None, Vespers, and Compline. Apart from the use of their voices in this practice the monks were expected to live a life of silence. In addition to the dispositional aspect of such a life, one cannot help reflecting on its "vibrational" aspect.

Down the left-hand side of the montage in figure 21.4 are images of standing waves (the zones of relative stillness) created by sound in air, fire, and earth. The top image shows the sound wave accompanying the passage of a bullet through air. Down from that is a standing wave created by a constant tone inside a flame. At the bottom we have a Chladni form, showing the standing waves caused in particles of fine sand on a plate whose edge is being played with the bow of a cello. On the right are six images of standing waves in water. The different images represent very shallow alterations of depth of focus in a single droplet that is being subjected to a constant sound frequency. There is a kind of three-dimensional mandala in the droplet. (Apart from the Chladni form, these are examples of Schlieren photography that can achieve the necessary focus variations.)

Another, more refined point about tonal vibration will be familiar to anyone who has taken part in sessions involving continual repetitions of a single note or cadence. This can apply to psalm singing, Indian puja recitations, Sufi zikr, Buddhist sutra chanting, and group mantra. When the gathering is "tuned," overtones are heard in the space above the group, in the same manner that individual mantra practitioners, Mongolian singers, and sonic enthusiasts are able to make overtones above the fundamental sound of their own voice, by manipulating the sound in the vocal chambers of the head. When a congregation achieves such attunement (and it will not happen unless there is a steady concord of sound) the whole environment can come alive with the effect. On occasion one hears the sound spontaneously build into layers of frequencies in harmonic relationship to each other.

The shape of the environment can greatly assist and intensify the resonances involved. There is a yoga center in Sussex (Hourne Farm, near Crowborough) with a converted cone-shaped oasthouse that does this remarkably. The bowl-shaped stone of Cistercian monastic architecture, unadorned and built to simple geometries—that already embodied musical ratios—was found to be ideal.

The subtle physical effect is that all the fluids, the breathed air, the heat fields, and, to some extent, even the tissues in the bodies of the people voicing the sound will be taking on standing waves. Our bodies are at least 65 percent water.

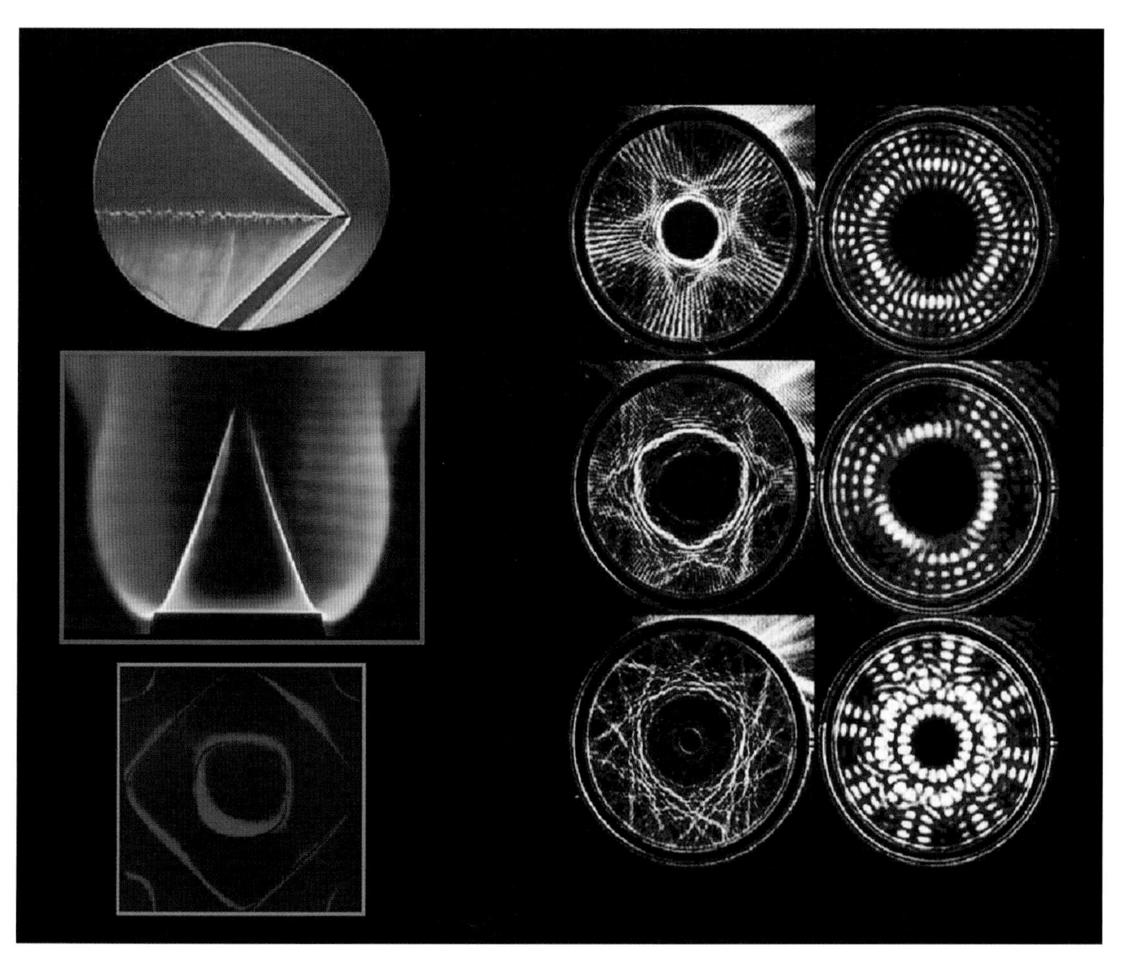

Figure 21.4
A montage of sound waves in air, fire, earth, and water.

So a Cistercian monk, eight times a day, for—on average—half an hour at a time, would sing communally in a bowl-vaulted chamber in precisely this way with the repeated notes and cadences of Gregorian psalmody, with that being pretty much the only use that he would make of his voice. No wonder the Cistercian movement had such power. Anyone who has taken part in well-tuned mantra sessions knows that, in the silence at the end of it, one feels extraordinarily centered, energized, and recollected. The physical aspect of that state of awareness is no doubt related to the persistence of standing waves in the system. It applies whether one is a Tibetan repeating one's mantra or a Kabbalist repeating the Divine Names or a Vedic practitioner chanting the ancient verses. Regular tuning, of this and other kinds, is fundamental to all effective spiritual practice.

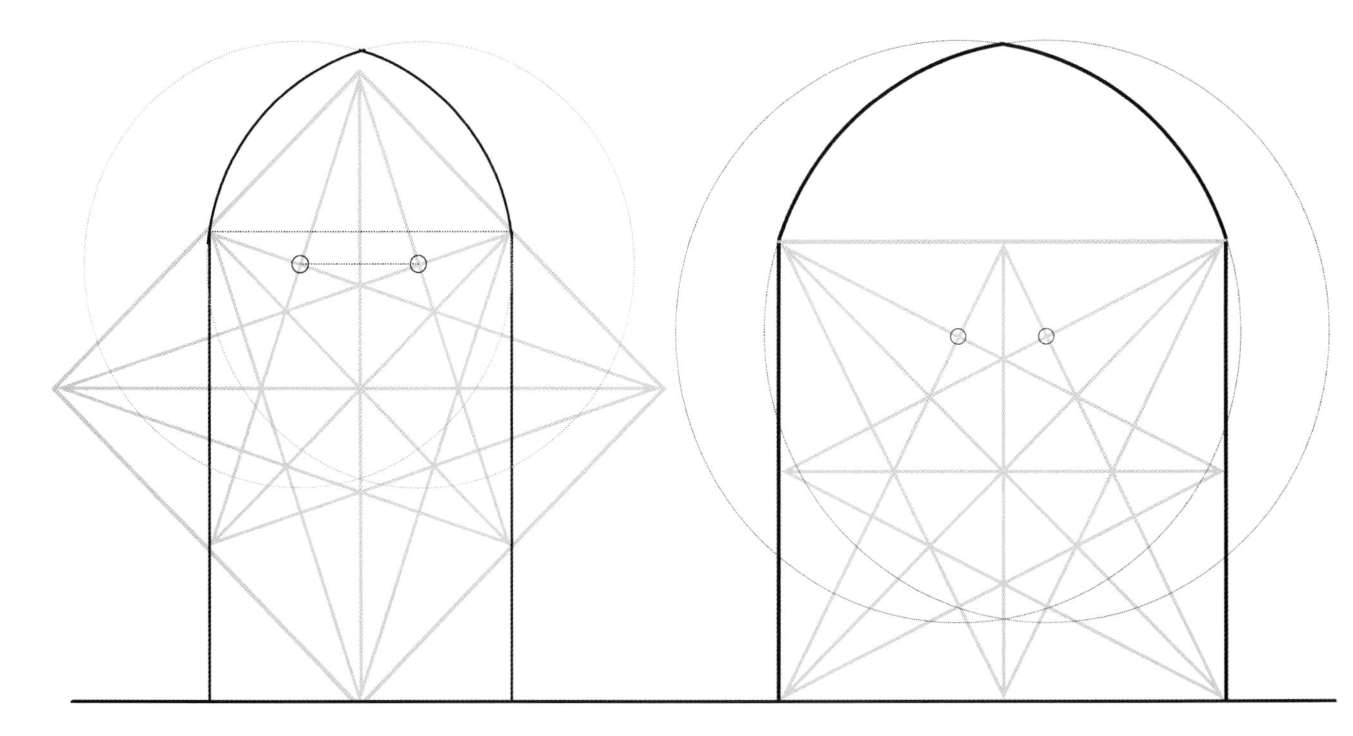

Figure 21.5

Typical Cistercian vault profiles. Wherever the base is set, in such a chamber the sound circulates as it reflects in the barrel shape, emphasizing both resonance and harmonic effects. The Starcut is a very useful matrix because it is both a flexible and harmonious design standard.

The great acoustician and hearing specialist Dr. Alfred Tomatis was once asked for his expert help in the case of a monastery at which the monks were becoming increasingly demoralized and unable to function. Spiritual exhortation, psychotherapy, vitamin pills, changes of work regime, and so on, had all been tried to no avail. It was a "modern" monastery that no longer required the monks to sing the Divine Office together. Having stayed there for a time and having observed this fact, Dr. Tomatis pointed out to the abbot that, before any considerations of "meaning," certain frequencies (between 2,000 and 6,000 hertz) that are abundantly present in the voice and its overtones first and foremost energize the brain. Through facile assumptions about religious modernity, they had switched off the community's main engine! They resumed the singing of the Office and their monastic life became livable again.

In terms of architecture, the lower measures of the chamber, beneath the bowl vault, may vary as in figure 21.5. All required proportions can be derived directly from the Starcut. The example on the left is very similar to those of the sanctuary and nave at Le Thoronet.

THE STARCUT AND APOLLO'S LYRE

> *Musical values introduce a hierarchy into the number field . . . musical patterns elevate certain numbers to a prominence pure number theory would not give them.*
>
> Ernest McClain

In the context of musical tuning, the mid-side diagonal, which has been such a feature of this whole study, has a further so far unmentioned quality that sums up its versatility.

In figure 21.7 (p. 231), by means of the vertical lines, a wide range of possible divisions are indicated. The white-ringed nodes on the matrix intersections mark the immediately visible divisions that generate the thin lines. The red rings mark further nodes where the thin lines cut the Starcut. To show what these divisions give us it will help to review some of the basics about the division of a musical string. There are

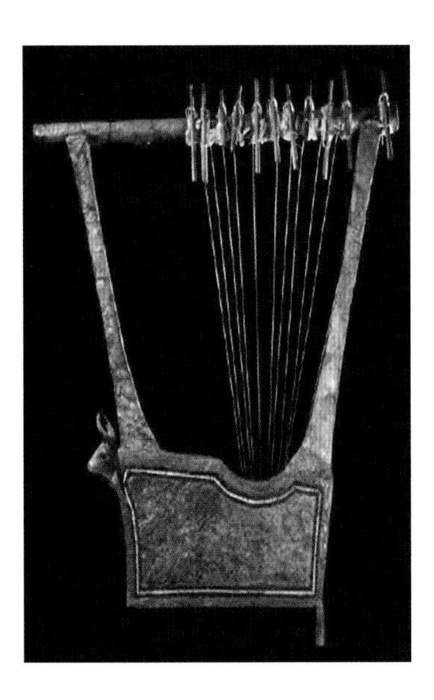

Figure 21.6
Sumerian lyre. Length and tension are what determine the pitch of a musical string. On this lyre strings can be moved laterally along the upper bar, changing their length and tension. They can be further adjusted by manipulating the pegs at the top of each string.

numerical ratios involved in what we hear as a melody. The delight of harmony is substance to substance-maintaining resonance.

In figures 21.7–21.9, the downward lines are all imagined as musical strings tuned to the same note. The Starcut provides the angle at which to set the bridge changing the string lengths of this virtual lyre, so that a rational and internally harmonious set of notes can be produced. They ascend from lower note to higher as one goes from left to right.

In the ratios that follow, the first number gives the measure of the full length of a musical string; the second number gives the amount of that measure that is actually sounding. After the ratio, I include the way it is written as a fraction.

The whole string is the fundamental 1:1 (1⁄1). The next longest length is the minor third that is 6:5 (5⁄6). The list follows (see also fig. 21.9):

Third 5:4	(4⁄5)
Fourth 4:3	(3⁄4)
Fifth 3:2	(2⁄3)
Minor sixth 8:5	(5⁄8)
Sixth 5:3	(3⁄5)
Octave 2:1	(1⁄2)

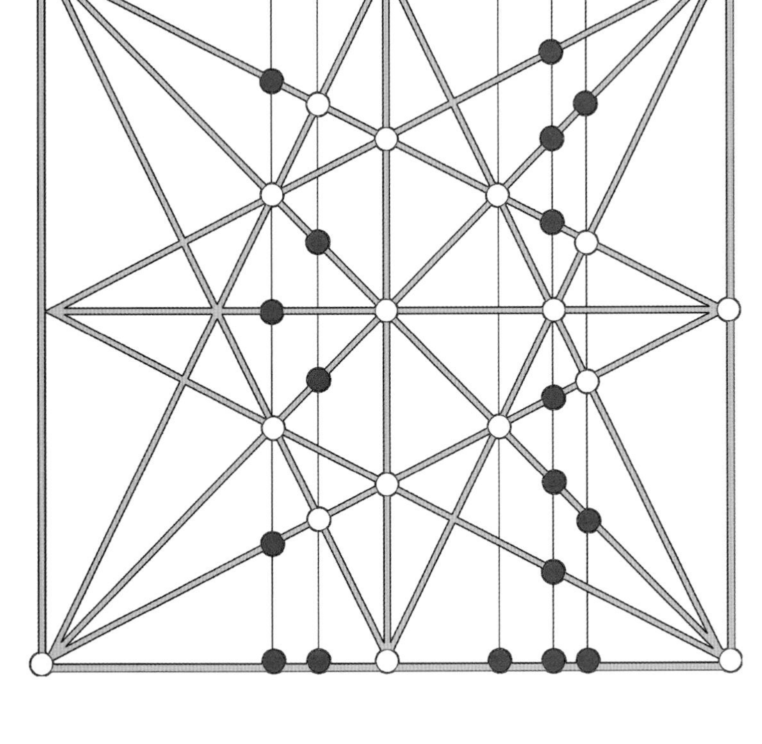

Figure 21.7
Division of lengths of strings.

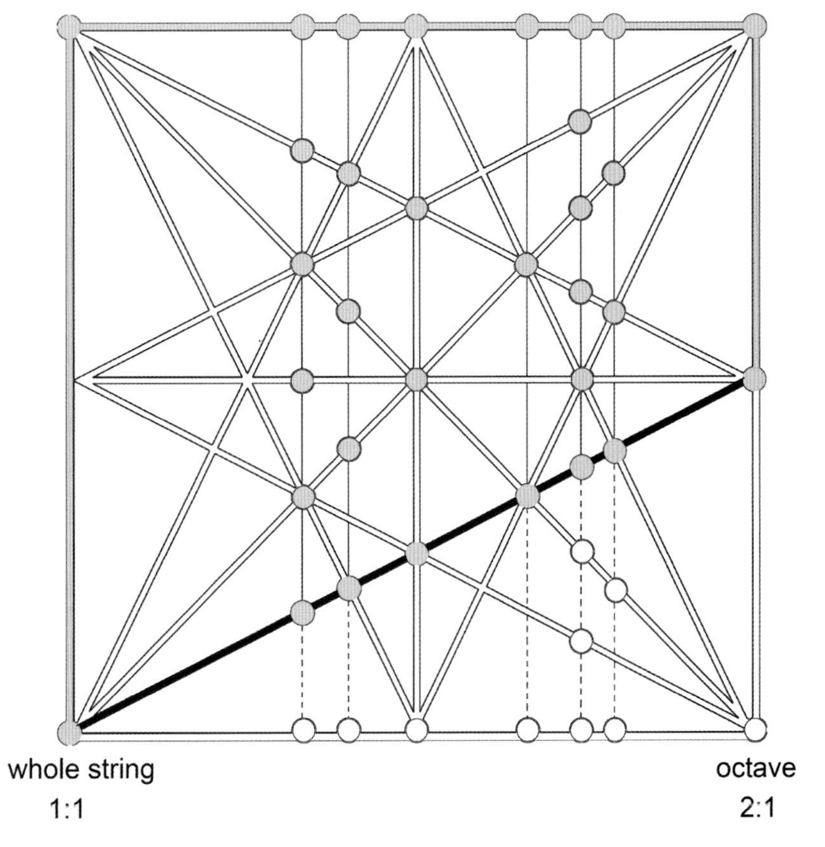

Figure 21.8
The bridge of the lyre is marked as a black line. A mid-side diagonal.

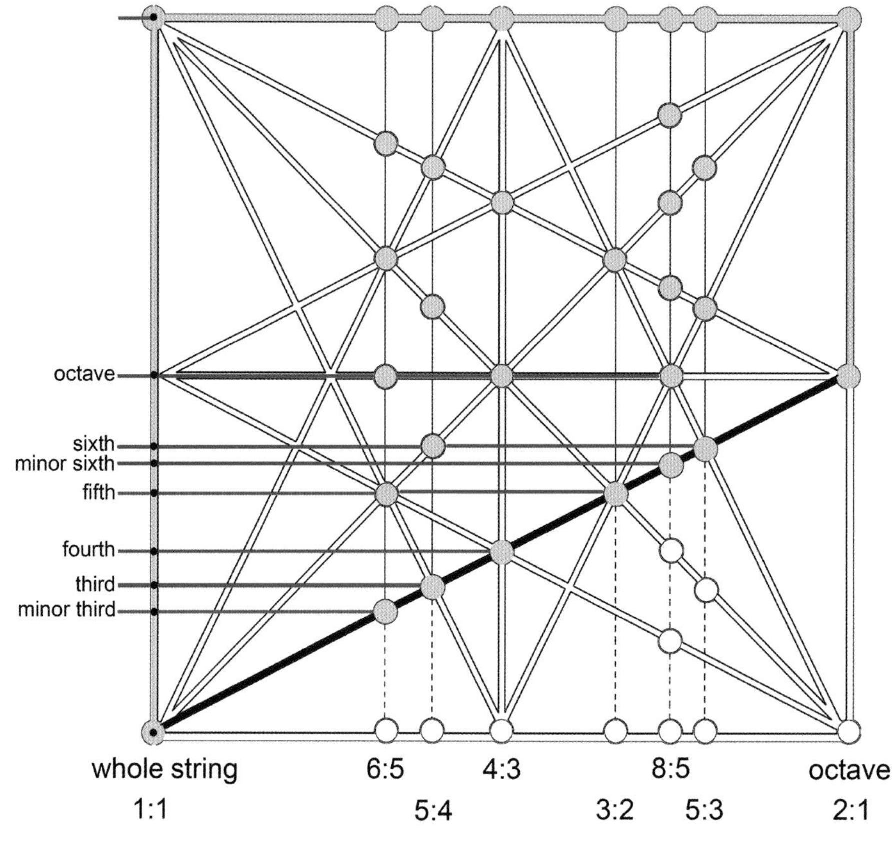

Figure 21.9

At the bottom are the string ratios. The red lines carry the lengths of the strings to a side bar that is marked with frets that would allow one to play all these notes on the fingerboard of a guitar.

A LAST HARMONY

Bearing in mind that the diagram contains its own fundamental "seed" of number in the area of 6 that is in the smallest 3-4-5 triangle (fig. 21.10) and that the overall area is thus 720, we get a beautiful set of musical values at the very center of the diagram (fig. 21.11).

The central octagon, in which the triangles are each of area 15, has a total area of 120 (fig. 21.11). It contains two squares of different sizes. We normally think of an octagon as being both of equal sides and angles. This one has equal sides but there are two different sets of angles—two underlying squares therefore. Their area can easily be calculated: 80 and 90 (figs. 21.12 and 21.13).

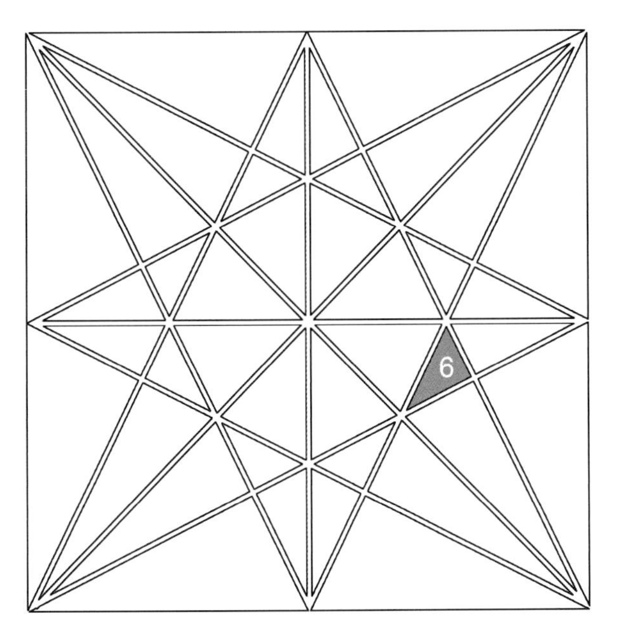

Figure 21.10

Returning to area 6, the smallest triangle.

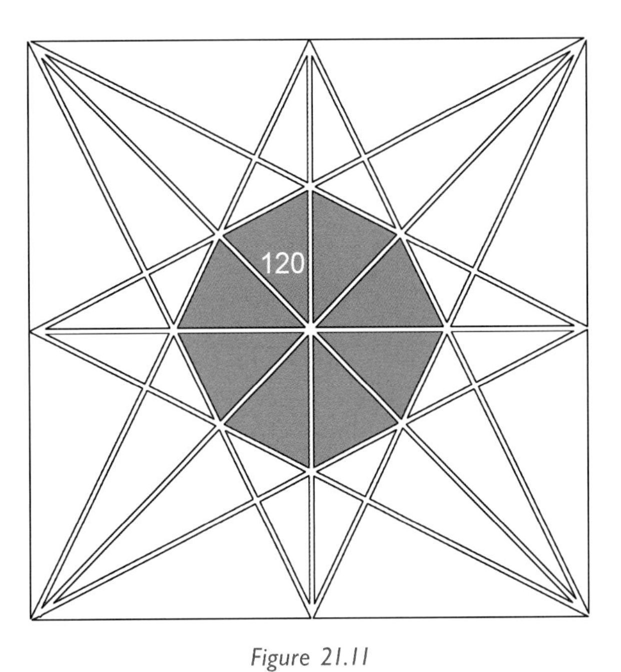

Figure 21.11

The "tone" at the center.

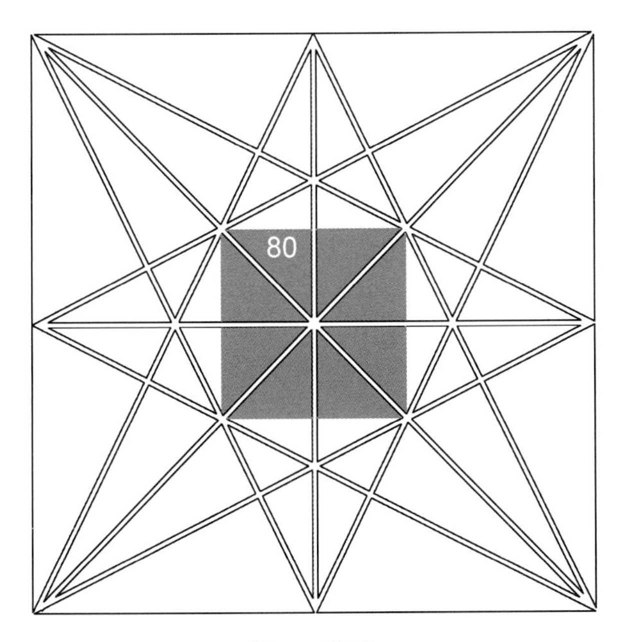

Figure 21.12

The 80 square.

Figure 21.13

The 90 "diamond."

As frequencies one can start with a pitch of 60 hertz or vibrations per second (fig. 21.14).

The perfect fourth above 60 hertz is 80 hertz (¾), and the full tone above that, which is the perfect fifth above the tonic, is 90 hertz, and the octave note is 120 hertz. Whether 9 × 80 or 8 × 90, both produce 720, the area of the entire square. The harmony of the numbers is as consistent, of course, as ever; and there is an extra beauty to be found in

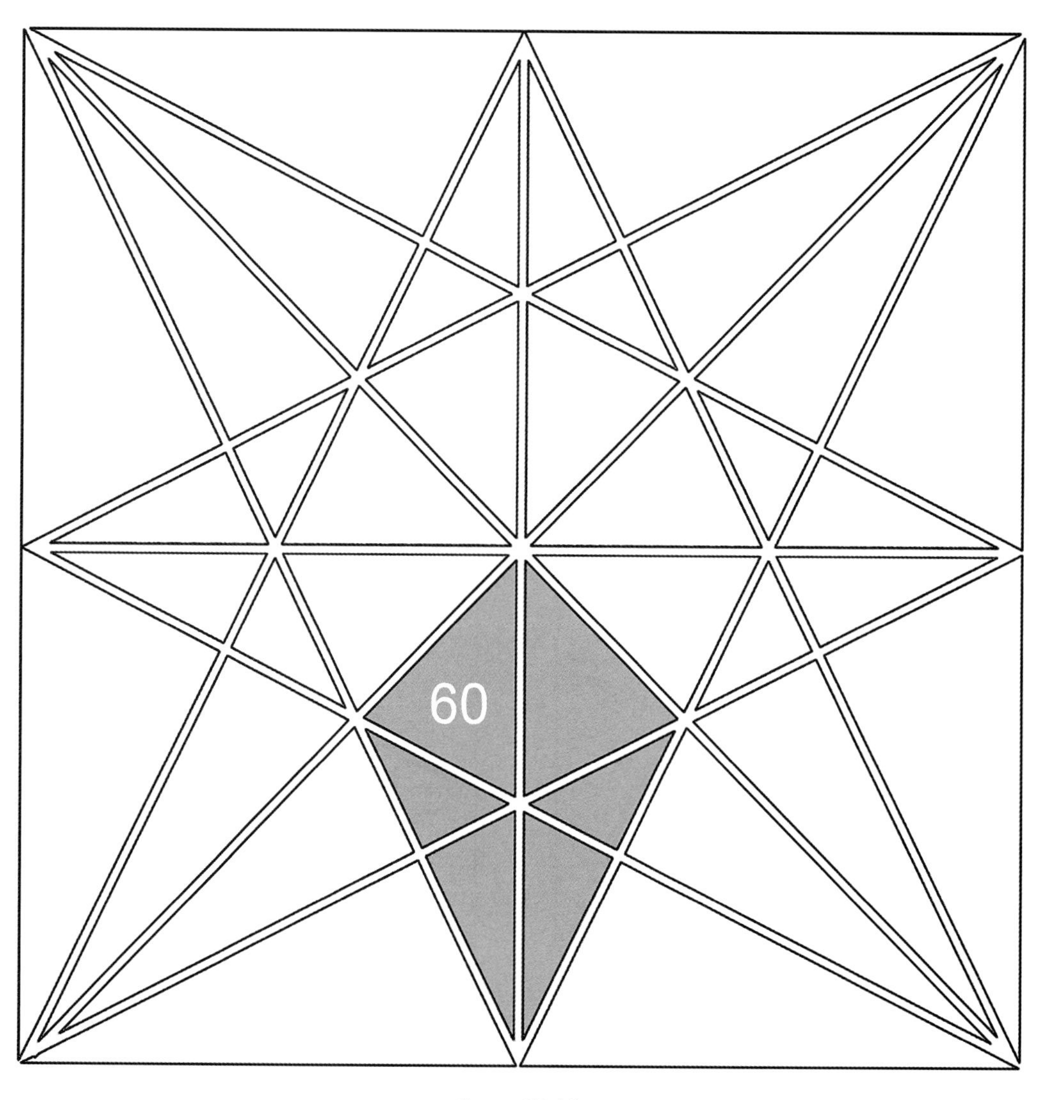

Figure 21.14
The starting pitch as an area of the Starcut.

the further octagon that only appears if one has drawn the interlocking squares. Its area is the number that has remained quiet ever since those original circles that generated the Starcut diagram in the first place—the 7. So where the squares that have areas of 80 and 90 (the full-tone relationship) overlap each other, there is hidden the same number as was hidden in the Bible's Temple measures, the number 70 (see figs. 21.15 and 15.2, p. 159).

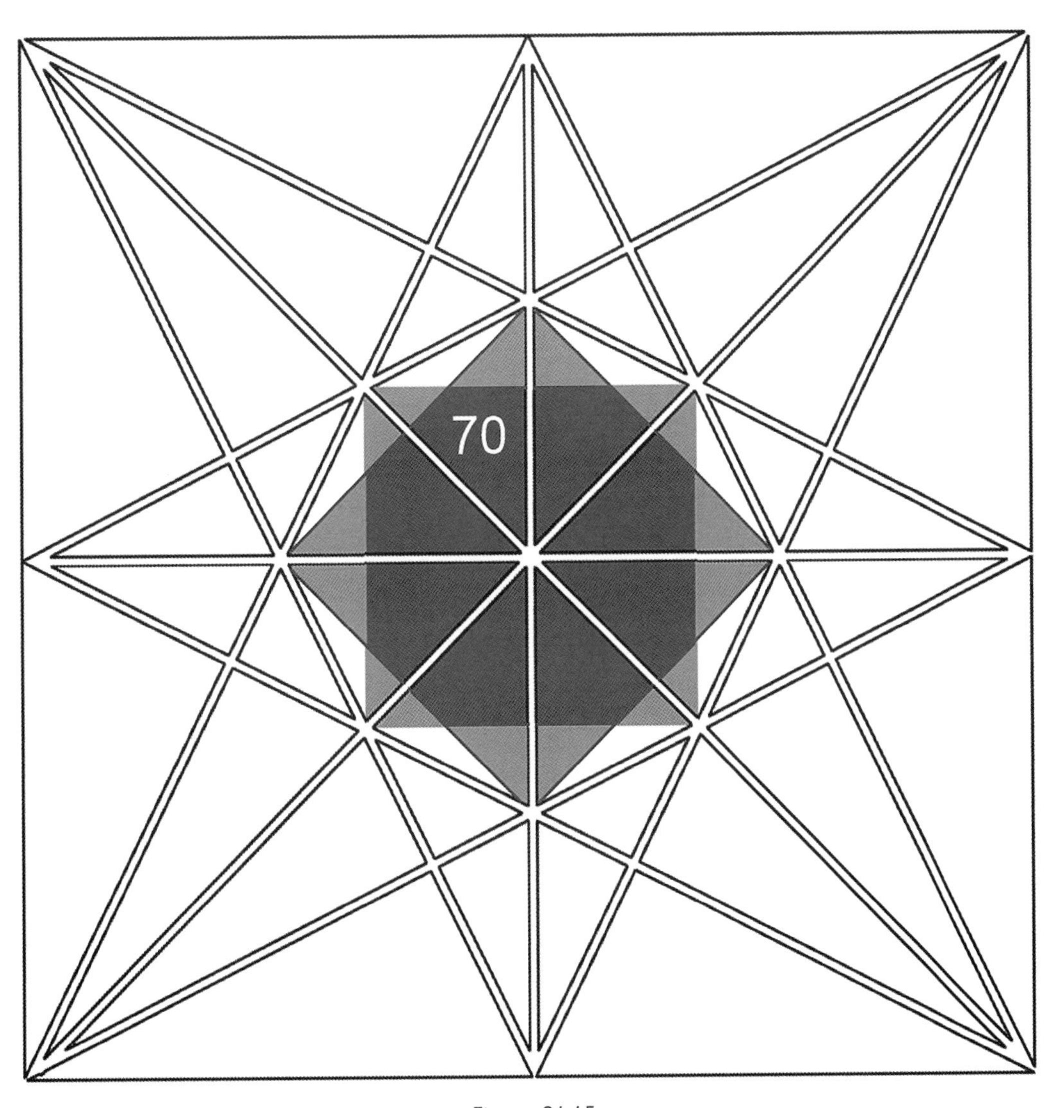

Figure 21.15
The shared area of 70.

Figure 22.1

Just discernible beneath a thousand-year-old text of Greek prayers there is an ancient copy of part of an essay by one of history's mathematical giants.

22
The Puzzle That Drives You Mad

Figure 22.2 shows three of the 536 possible solutions (see Lahanas, who credits Cutler 2003) to a puzzle that was for well over a thousand years only a rumor mentioned in ancient writings until, early in the twentieth century, in a monastery collection of old documents, a researcher noticed the faint traces of a previous text that had been scraped off so as to be overwritten with Greek Christian prayers. Then the thing got lost for another eighty years or so, turning up again in 1998.

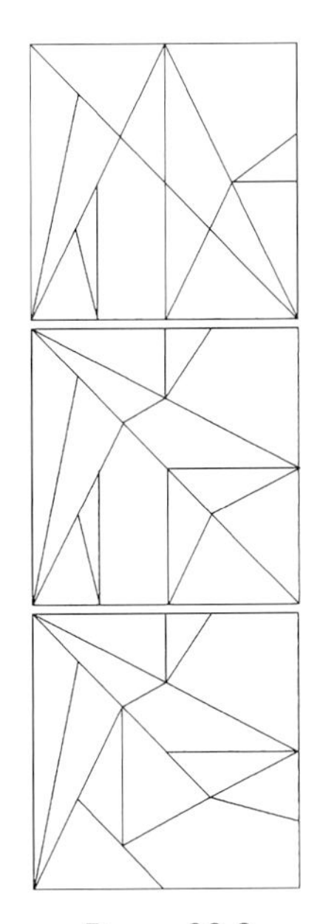

Figure 22.2
Some solutions to the "battle of the bones."

Thereafter this palimpsest, which is what such a thing is called, was deciphered. It turned out to be a copy of an essay by none other than Archimedes himself.

Some mathematicians were at first disappointed to discover that the material was devoted to a puzzle, known nowadays as the *stomachion,* said to derive from a Greek word meaning "stomach" or "container." In Latin literature it was referred to as the *loculus,* a word with a similar meaning. The name presumably referred to the fact that the fourteen separate pieces, often made of ivory, had to be kept in a receptacle. Some researchers prefer a reading of the name that comes from a fourth-century document, suggesting a different meaning: "the battle of the bones."

As can be seen in figure 22.3, the pieces were all geometrically shaped slices of a square, and the game was to recombine them into differently arranged squares or into the forms of humans, animals, or objects. It was considered a children's game, but according to some accounts it had a reputation for driving the player crazy.

Did Archimedes invent this puzzle? Expert opinion doubts it, regarding its origin as earlier and unknown. But if he did not invent it, why was he interested in it at all? His fame is for wide-ranging mathematical calculations such as his "sand reckoner" proving that the seemingly innumerable grains on the sea shore were, in principle, countable; for his figuring out of the displacement of fluids; for solving pressing real-world problems such as how to destroy the Roman galleys besieging his home city of Syracuse, by setting fire to them with focused sunlight from concave mirrors, or by capsizing them with hooks lowered from the shore battlements to snag into their lower hulls so that they could be upended with winches. Why, modern mathematicians wondered, was such a man interested in a child's game? However, those mathematicians who thought they would find all possible square solutions in a

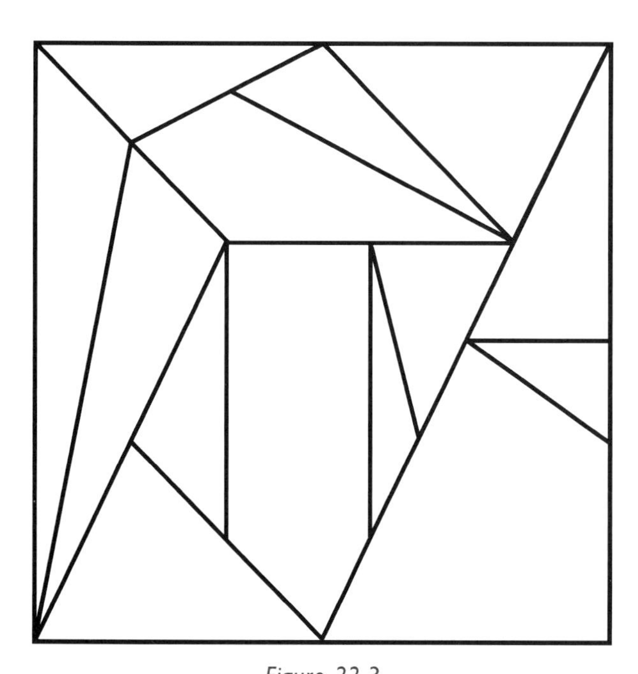

Figure 22.3
The slices of a square.

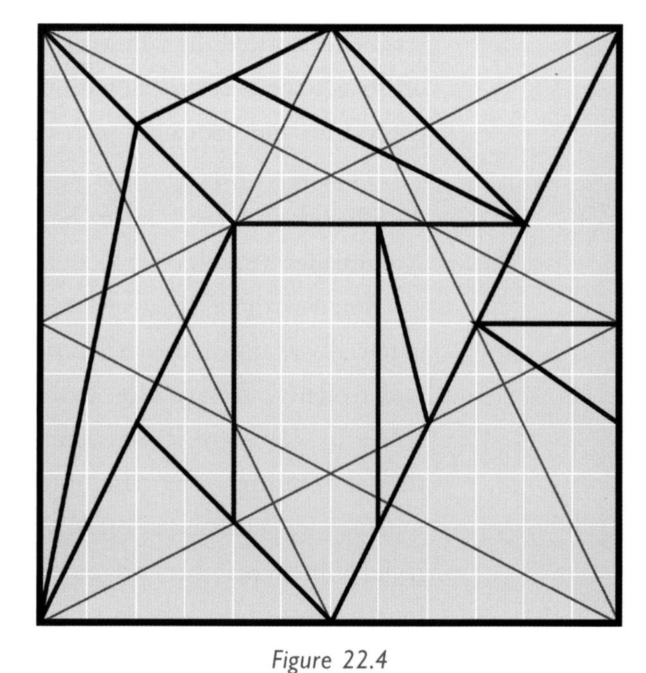

Figure 22.4
The 12-grid and Starcut overlay.

couple of hours, were chastened to discover that it was really quite a challenge and some interesting new math grew out of their—finally successful—efforts.

The observation that follows here proves nothing; but it does hint at the Starcut as a possible source. Whoever invented the thing, if familiar with the Starcut diagram, might well have spotted that it could easily be adapted into the maddening puzzle presented by the fourteen shapes.

Playing around with the pieces one soon sees that certain gradients (slopes) dominate. Of the fourteen pieces, only three include a single gradient that is not in the Starcut, and all can easily be derived from it. The rationale for the puzzle is patterned on a grid of 12 × 12, which can be produced with ease from the Starcut, as in other ways of course. When, however, that grid and a Starcut are overlaid on the pieces many correspondences become obvious. With the square's area taken as the standard 720, all the areas of the *stomachion* are multiples of 15—nice! (See figs. 22.4 and 22.5.)

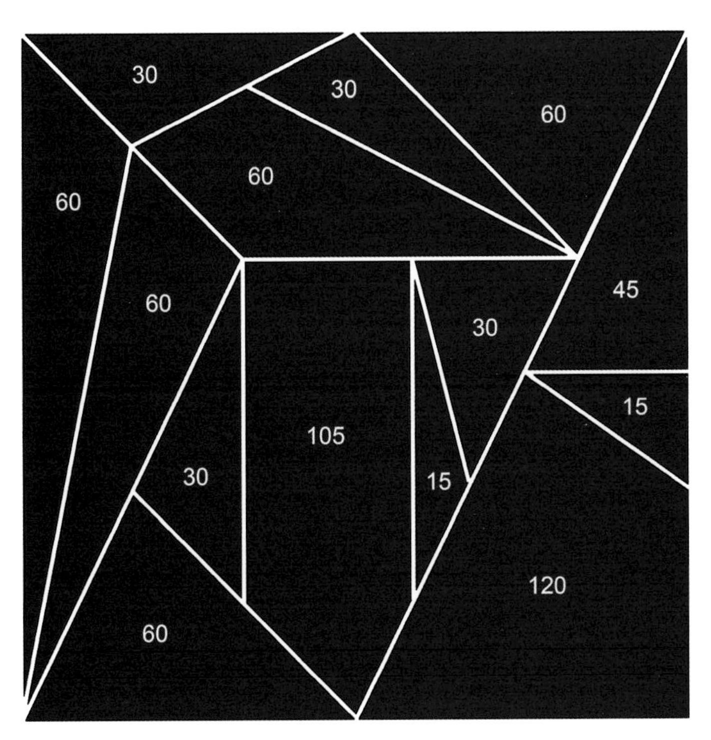

Figure 22.5
The areas of the pieces.

23
Reflections within the Light

An analogy for the graduated order of things in the world of the senses can be gathered by someone seeing the light of the moon coming in through the window of a house, falling on a wall mirror from which it reflects onto another wall whence it falls and illuminates the floor. So the light on the floor is from the wall, which has it from the mirror, and the mirror has it from the moon, but the moon's light comes from the sun. . . . So are the lights of the heavenly realms ordered such that the highest is that which takes its radiance from the Ultimate Light.

AL-GHAZALI, *NICHE FOR LIGHTS*

Opposite: Figure 23.1 Sunrise, Garry Beach, Lewis, Scotland.

Such has been the meandering nature of the research leading to this book that the image in figure 23.2 (p. 243) was unknown to me until this study was all but completed. It has long been familiar to scholars of the alchemical tradition, but the Starcut image within it has not occasioned much curiosity. It is, however, of great interest when taken in conjunction with actual alchemical practice and with the theme of mediation (geometric, arithmetic, harmonic, and "golden") that we have touched upon in connection with Greek cosmology. The same theme is implied in the notion of the declination of light given by the Sufi theologian Al-Ghazali quoted above.

The true purpose of alchemy has been described by a contemporary alchemical philosopher as "the realization of the internal golden essence"—meaning the transformation of a person's ordinary life-energies and awareness into a higher spiritual potential that is, in fact, already innate, but is obscured by the natural accumulations from earthly conditions. The fires of alchemy are the processes whereby that dross is transmuted. Such processes work toward the refinement and redirection of sexual vitality, the clarification of the thinking, and a spiritualized opening of the heart.

For the Christian alchemists of the Renaissance, of whom Stephan Michelspacher seems to have been one, that which vivifies this Great Work is "the Light" that is mediated from its Divine source both via the Christ and the natural "lights" of the celestial bodies. The particular form of the Starcut in Michelspacher's *Mirror* emphasizes this imagery by picking up on light's catoptric quality, which is precisely the way it reflects as from a mirror.

This alchemical agent—"Light"—remains an enigma to those who try to understand alchemy from a solely theoretical point of view. For what exactly is meant by "Light" in this context? Is it simply a figure of speech? Is it some mystical radiance? Is it natural light? Is it the light of imagination?

First, "Light" here is far more than merely a figure of speech. The philosopher Elémire Zolla hinted at a dimension of light that we rarely consider. I can no longer trace the exact quotation from his paper *The Uses of Imagination and the Decline of the West,* but I well remember the question he posed: When we are asleep in the dark at night, with our eyes closed, and we see things happening in a dream . . . where does that light itself, by which our dreaming mind is actually seeing, come from? Light, since Zoroaster and perhaps before him, is Divine; and light is directly associated with consciousness itself.

Second, it is known that "SAD syndrome" depression can be cured by exposure to the light of a lamp, immediately followed by a

Figure 23.2

Illustration from *Kabbalah: Mirror of Art and Nature,* published by Stephan Michelspacher in the early seventeenth century. At the upper left within the general heading "Natural Matter," are the words: "Kabbalah and the Alchemy give you the highest medicine and in addition the white stone, in which alone lies the foundation, as will, at the appointed time, be revealed through these figures." (The white stone, mentioned in the book of Revelation, is given to each of the redeemed, bearing a name that only they shall know. I take this to mean a person's true inner nature. The phoenix is the emblem of regeneration.) At the upper right, labeled "Ultimate Matter," are the words: "Oh God, help us to be thankful for this gift, so exalted and pure; when you open the perfected inner heart and senses to accomplish this work shall you have all strength." (The lion is the symbol of the transformed inner forces.)

period of eyes-closed darkness wherein the person concentrates on the afterimage of that light. It is a useful healing technique, but it is only the tiny peak of a veritable mountain of therapeutic, psychological, and spiritual knowledge developed from the nineteen sixties through to the eighties by Professor Francis Lefebure, the eminent French ophthalmologist. His pedagogical work with children and adults with special needs was as effective as it was groundbreaking. His book *Phosphenism: The Art of Visualisation Developing Memory and Intelligence,* is hard to find but worth the search. His claims for the beneficial, consciousness-enhancing and mood-transforming effects of what he called phosphenic mixing were backed up by decades of recorded practice. What emerges as one studies his work goes beyond therapy into the positive psychology of joy and indeed into visionary mysticism. Who tastes knows.

Third, the philosopher Oscar Ichazo teaches methods of energy generation, dissolving of conditioning, imaginative visualization, and the uses of enhanced natural light to achieve the "state of transcendence"—the culmination of what I have called, elsewhere in this book, the ancient path of return.

Unfortunately the image in figure 23.3 is of low resolution, but one can see how the light ray issues from the divine Tetragrammaton (the four-letter name) in the upper left, and passes through the Hermes figure on the top of the fountain. Hermes is the divine messenger. This association is emphasized by the Holy Spirit as a dove touching the (unfortunately indistinct) crucifixion scene in Jerusalem at the upper right. There the light sweeps in an arc to the central crowned Christ figure who offers chalices of divinized nature to the realized sun/moon figures. They, as sulphur and mercury, are the soul and spirit aspects of the alchemist. The Christ figure is also pictured in direct ray contact with the Divine. The angle made by the arcing rays is exactly the corner angle of the Starcut, shown in full in figure 23.2.

It is in this figure that we see the Starcut as a focal image of the Great Work. The creative solar light of the central circle reflects catoptrically in all directions throughout the field of the other planets and thence into, and beyond, the crosses at the four corners denoting this world and the alchemists' transformed human nature.

It was satisfying to come across this use of the Starcut with a directly transformational association. It is tempting to unravel all of the symbolism, but that is beyond our present scope.

Figure 23.3

Upper detail from the fourth engraving in Stephan Michelspacher's work. It shows the motion of the divine ray.

Figure 23.4

Zodiac cartoon for an engraved mirror. The stylized winged forms spanning the corners symbolize "breaths." In Sufi tradition a complete breath is denoted by a star octagon (an in-breath) within an open octagon (the out-breath). The twelvefold wheel of time and qualities has an inner calibration of sixty, as it does in the Mesopotamian number system that we still use. The central device about the unity-point denotes the "twenty-four Elders around the throne" in the book of Revelation or, with the further twelve points that look inward, the thirty-six decans of the 360° circle. Thirty-six degrees also refers to the pentagon star as the angle at its points. Ten radials, thirty-six degrees apart around the center of a circle, will make the points of a decagon on the circumference. The numbers embodied in the central area symbolize the unchanging Eternal. The outer area of the panel is a visual metaphor for the wheel of life within time.

Figure 24.1

Salim Chishti's Tomb at Fatehpur Sikri, near Agra, India.
Salim Chisti (1478–1572) was a greatly revered Sufi mystic who, it was thought by many, could perform miracles.

24
Inevitable Sufis

I sensed that . . . the zone of experience was not chaotic but ordered, and that, as our consciousness cleared, we actually participated in the unifying process of actions. If madness was the conviction that there was some total conspiracy to imprison and destroy us, perhaps the wisdom of the sages was that, in fact, this was but the dark shadow of a luminous truth—that the world was a divine conspiracy to liberate and re-create us.

IAN DALLAS, *THE BOOK OF STRANGERS*

One of the great, and only more recently appreciated, influences on Western philosophy, art, and spirituality is undoubtedly the high cultural tradition that flourished in Islam, though not without opposition, under the name Tasawwuf—the "Sufi" stream. I first encountered its wisdom in 1962 in the teachings of Hazrat Inayat Khan. Thereafter in the mid-1960s the works of Idries Shah began to appear, and in some haunted way one felt attracted to seek a "golden road" to a rumored haven of the soul. It was while living in Sri Lanka in 1972, and in a more mundane context that I first came directly under this cultural influence. The *Colombo Times* reported on an educational development in Britain using new combinations of Arab polygonal patterns, only then possible due to computers (magical beings in those days). The development was ascribed to mathematician Ensor Holiday and helpers, and, one gathered, the project had been initiated by Idries Shah. The paper reported that the concentration span of children with attention deficit disorders had been enormously improved, and their creativity greatly enhanced, when they colored in these patterns. Some months later on my return to the UK, I found that what had been welcome as far away as Sri Lanka seemed to have gone unnoticed on its own home territory; it was only after weeks of asking in mainstream stationers and toy shops that I tracked down the patterns in a small specialist bookshop in Edinburgh. They were called Altair Designs—now they are commonplace. Figure 24.2 is an example of the latest appearance of these patterns in John Martineau's *Altair Raindrops*.

It was in illustrating these children's pattern pages that I first developed an interest in the Arabian tiling tradition that was later useful in my glass mosaic work. In Altair there was a subtle reordering of two-dimensional space to allow "impossible" geometries where five-, six-, seven-, and eight-pointed stars and polygons could all somehow integrate on one surface. There was a psychological analogy here with the astringent and paradigm-changing psychological methods of the Sufi lore that I picked up through reading and contact with Idries Shah and others. Amid the new cultural impacts, hardly noticed by me at the time, was a particular page of a book called *Secret Societies,* ostensibly by someone called Arkon Daraul (whom, then and to this day, I have assumed to be Idries himself—a regular user of noms de plume).

The page showed a detail from a Muslim heraldic coat of arms of the traditional sheiks of Paghman in Afghanistan, the Musa Kazim family—a central branch of whom are the Sayyeds ("descendants of the Prophet") of whom Idries was himself one (fig. 24.3, p. 250).

At the center is an eagle on whose head is a plumed crown, making it look a bit like a hoopoe (a Sufi symbol), and on whose breast is a regular octagonal device that Idries used, as is still used, as the colophon of his Octagon Press. The star Altair, incidentally, is in the constellation of Aquila, the Eagle.

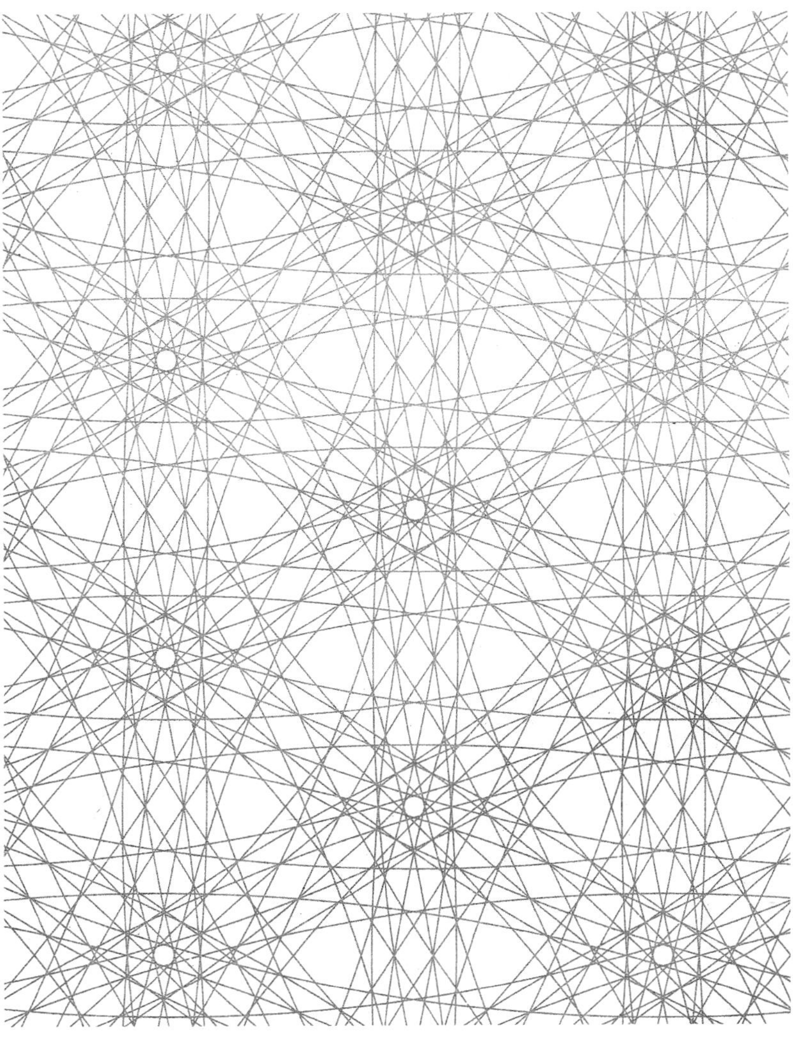

Figure 24.2
Image from *Altair Raindrops.*

It was only relatively recently that I paid attention to something about that image that had always nagged: the rather ugly (as I thought) crenellated, black bands surrounding the central figure. That semi-regular octagon . . . Then it clicked, and I superimposed the heraldry on the Starcut (fig. 24.4, next page). Their connection is obvious, and yet remains mysterious.

As elsewhere the provenance, or route of passage, from earlier times into a later—in this case Muslim—tradition is unclear. There

Figure 24.3
The central device of the Paghman heraldic arms.

Figure 24.4
Coincidence? The semi-regular octagon with the Starcut superimposed.

seems no problem in terms of general plausibility, but the finer detail of connection is elusive. There is clear cultural evidence of a deliberate integration of Islam with the classical tradition in the efforts of the Ikwan i-Safa—the Brethren of Sincerity. They were a group of Shia scholars of the tenth century in Basra who endeavored to put together an over fifty-volume encyclopedia of all that was then known. In it they sought to accord the Greek thought of Plato and Aristotle with Muslim faith. Greek culture provided the foundation of their science. A century before Islam, in 529 CE, the Near East had already absorbed much from the Hellenic world through the offer of refuge to Damasius and his exiled fellow academicians, by King Khosrow I of Persia, when the emperor Justinian froze the funds of all non-Christian institutions, and the Platonic Academy was disbanded.

While the above suggests possible connecting routes for geometric esoterica, what seems strange is that with all the Arabian development of geometry, and geometric mosaic and tile work, the Starcut lattice hardly ever appears. I have searched hard and long but have seen only one example, sent to me by a Muslim interested in my work. It was on the frontispiece of a fourteenth-century Koran. Overall within Muslim and Sufi motifs, as with the Starcut in the Renaissance, there is a tantalizing presence within a general silence.

I know of only one other Sufi usage, and it was recent. I noticed it, as a framed drawing on the wall of the meditation room in the Surrey household of the late Fazal Inayat Khan of the Chishti Sufi lineage—the lineage that had been brought to the West, in a form available to non-Muslims, in the early twentieth century, by the Indian musician and sage Hazrat Inayat Khan. Particularly in the Chishti, though also in other Sufi traditions, music is used as a gateway to the ecstasy known as *fana*—annihilation—where the sense of self is transcended in an experience of Unity. Fazal Inayat Khan at one stage used the Starcut as a working device for observing the activities, faculties, and "context" of the "Self" and the "Overself." As an educational, semi-yantric mnemonic it seems quite a sophisticated tool (see fig. 24.5, next page).

Apart from the universal concepts that are incorporated into it, and whose interplay is indicated by its pattern, it was the sheer occurrence of this geometric figure that intrigued me. It is possible that there is some prior history of its use for mapping specific experience; Fazal elsewhere made points that seem apt:

> As best as we can tell, in the ancient form—the inner line—of Sufism . . . they united to study three things: numerology,

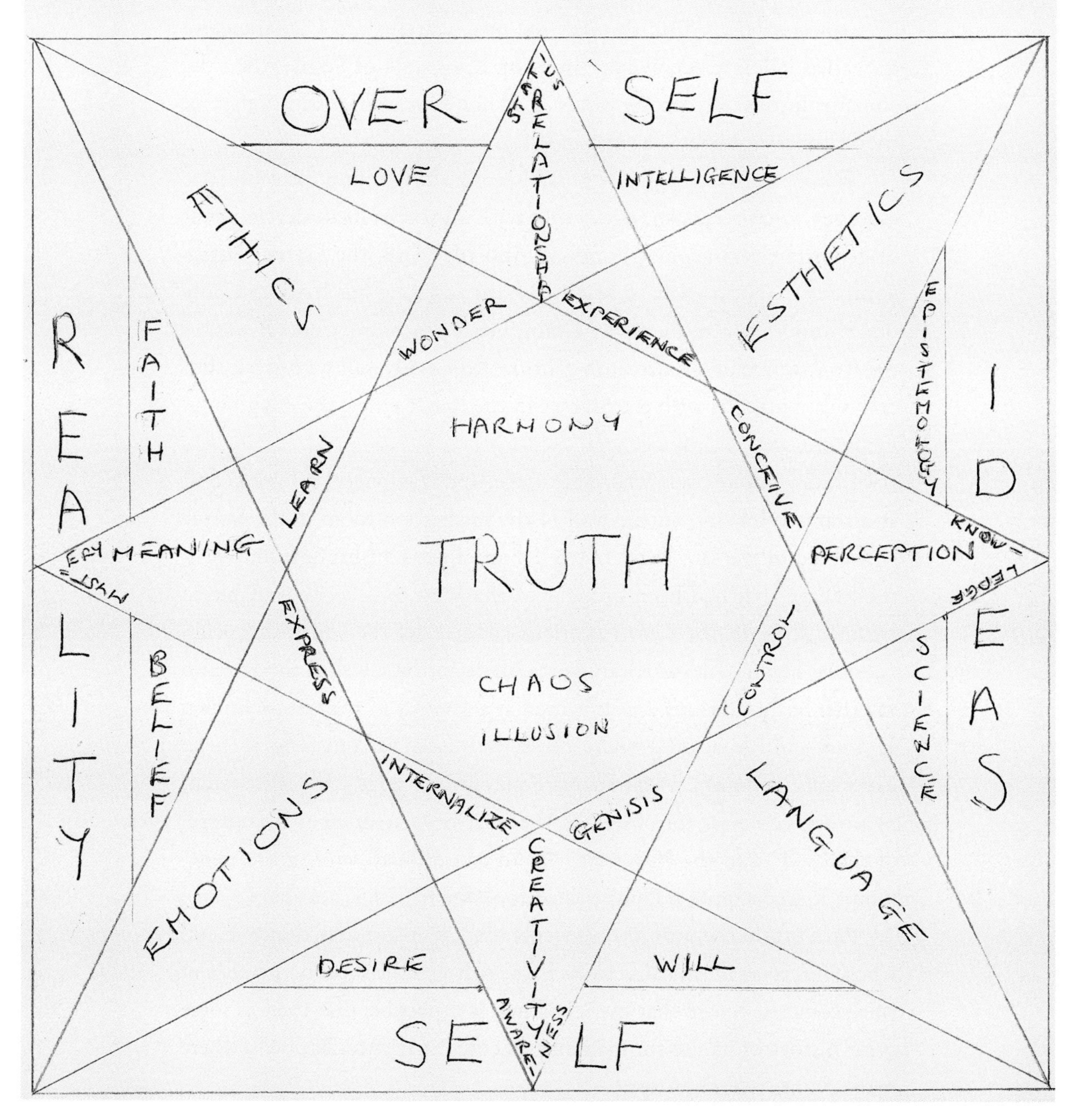

Figure 24.5

A by-hand copy given to the author by one of the late Fazal Inayat Khan's friends.

> symbology, astrology—these were the three basic researches of Sufism. Everywhere, in Christian mysticism, or Jewish mysticism or Hindu mysticism, these three will crop up again: numerology is the science of vibrations, of patterns; symbology is the science of forms, structures; astrology is the science of influences, psychology.
>
> Fazal Inayat Khan,
> "Non-Islamic Sufism" (essay)

There is a magic-square tradition in Islam, and magic squares may well have traveled westward along the silk, spice, and incense route that comprised a network all the way from China to Byzantium, to Mecca and to the caravan roads of western Asia. The relevance of the magic square to the Starcut diagram will emerge in the next chapter. A famous Sufi adage says: "Seek knowledge first in China." I did so long ago, but have taken until the writing of this book, and the chance visit of a friend, to understand more about the seminal patterns that are to be found there.

Figure 25.1
Day of the Yellow Ancestor
Beside the Yellow River.
Knowledge, ghosts
Of certainty; ours
A matter of chance.
John Esam, "The Yellow Ancestor"

25
The Luo Shu and the Limping Dance

"Looking at those eight triangles in the center, each with an area of fifteen," said the monk, "one can't help thinking of a Chinese legend from four thousand years ago."

The Chinese legend was that Fu Xi, the first of the three mythical ancestors, was sitting one day by the Luo, a tributary of the Yellow River. A turtle or "dragon-horse" emerged and, in the markings on its back, he made out a mysterious arrangement suggesting numbers (see the right side of fig. 25.2, p. 256). He copied the markings and initiated thereby the art of calligraphy. The array that he copied became the primary magic square at the deepest numerical core of Chinese philosophy. It determined one of the classical layouts of the Pa Kua—the eight trigrams of the Chinese oracle of changes, the *I Ching*. It was incorporated into the shaman's dance known as the "star walk"—or the Yu Bu. The underlying schema of all these are, in essence, identical to the Starcut, as we shall see. First, however, in figure 25.3, we see the traditional Taoist depiction of the numbers that the legend records as being on the turtle's back.

Figure 25.2
Fu Xi and the Yellow River diagram.

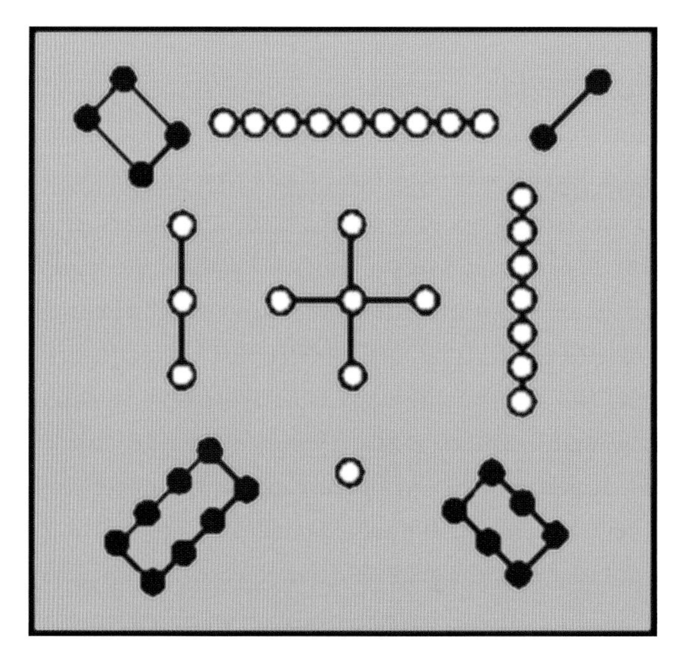

Figure 25.3
The "Luo Shu" or Yellow River diagram.
Odd numbers (yang) are white; even numbers (yin) are black.

If the numbers are translated into our number symbols we find that they make the magic square, known in the West since medieval times, as the square of Saturn (see fig. 25.4).

In this ancient arrangement the numbers add to 15 in all directions. Including reflections, the integers can be arrayed in eight ways without altering the magic quality. Five is always central. The square has other features, of which more later.

The Pa Kua in figure 25.5 is in the "later heaven" arrangement of the Chinese cosmos. It is laid out according to traditional associations with the numbers of the Yellow River diagram. The yin-yang symbol in the center is considered as denoting twofold polarity, and it does so. It does, however, have a fivefold quality, appropriate to its position in the middle of the square: there are two cells, two seeds-of-the-other within those cells, and one unifying circle. For me this fivefold-ness is also a mnemonic to recall that for the Chinese there were five elements in play. The order of the creative cycle of elements, known as *sheng,* was earth, metal, water, wood, and fire (fig. 25.6, next page).

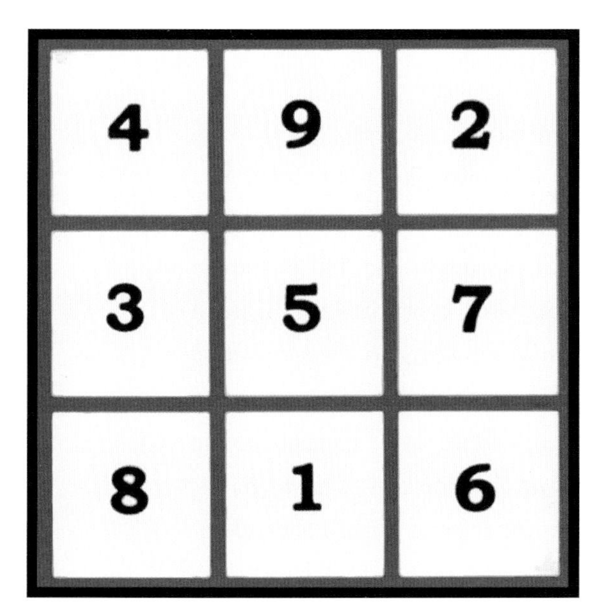

Figure 25.4
This is known as the basic "order three" magic square.

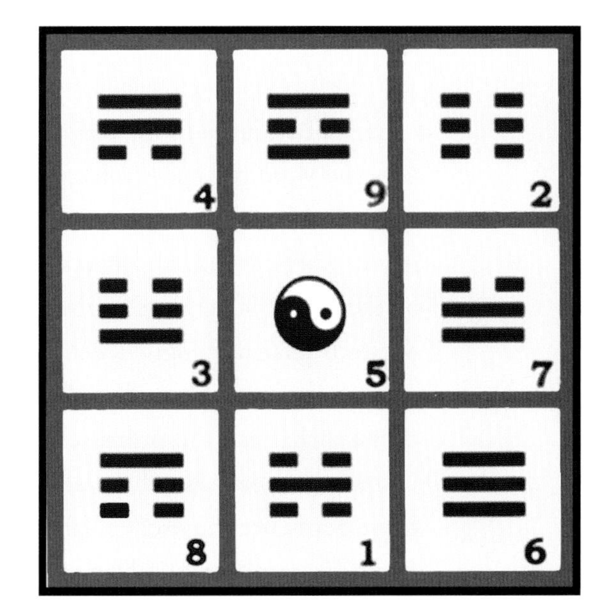

Figure 25.5
The trigrams, from combinations of which are formed the hexagrams of the *Book of Changes,* are arranged here in a form called Pa Kua.

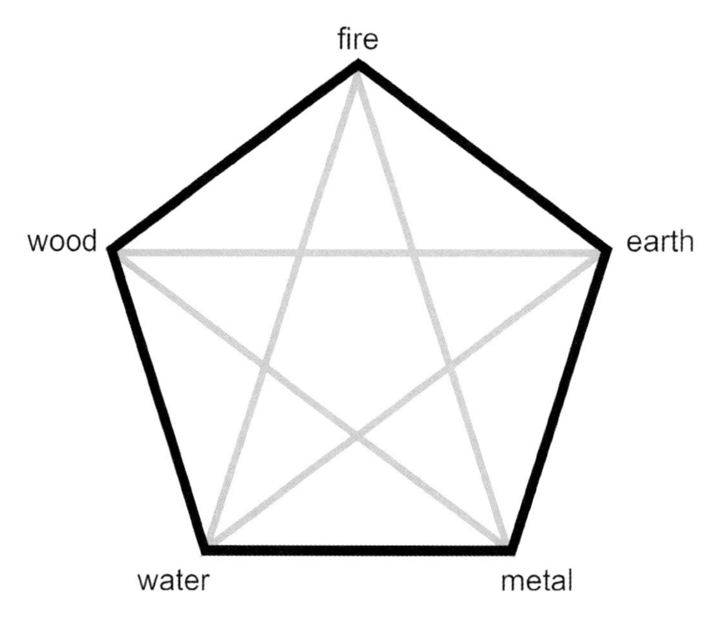

Figure 25.6
The pentagonal arrangement of the elements emphasizes that their archetypal order was as a cycle, not an end-stopped line. There are internal movements in the pattern that are relevant to the Chinese healing arts.

Lars Berglund, in his very informative work *The Secret of Luo Shu,* uses the word *phases* in reference to the elements. This is helpful as a reminder that the complex process-states of matter embodied in earth, metal, water, wood, and fire are taken as phases in the manifestation of the universal chi—the energetic ground of everything. I heard a Western pharmacologist recently ignorantly referring to these Chinese concepts as mumbo jumbo. If one spends a few moments reflecting on just what are the processes in these different states of matter, the Chinese approach is subtle, perceptive, and informative. These phases, or elements, are themselves balanced as number pairs within the Yellow River diagram (the Luo Shu).

Five, at the center, was the focus of Chinese culture. But it too was poised in the balance of forces that were expressed in the yin and yang. In figure 25.3, the Luo Shu graphic, notice that the cross of yang numbers (white) is boxed by a square of yin numbers (black).

The yin numbers are at an angle outward/inward and mark the intermediate directions, as for instance northeast where the two is situated. On the diagonals, the yin pairs each add to ten as do the yang pairs in a count where the central five was ignored. The diagram itself is in the balance of yin and yang that it symbolizes. The Yellow River diagram, because boxed inside the square of the even, yin numbers, was considered to be, in essence, yin. It was the mother of the entire Chinese number system, popular aspects of which were to flourish in a vast lore of Taoist geomantic and talismanic magic. The diagram could be laid out with knotted string, and the even numbers tied so as to loop conveniently. Tying knots has been a mnemonic and magical practice in a number of cultures. One only has to look at Celtic decorative knot-work to appreciate that it flourished in the West as well. Certainly wherever there was a maritime or fishing culture, knotting was known. In legend Fu Xi first taught the people to fish.

SNAKES AND LADDERS

I think magic squares, like much of what has been discussed in this book, may have started as graphic arrangements that acquired their numbers because of the simple enumerability of their patterns. Certainly it is unlikely that magic squares came about without people having noticed things about more ordinary arrays of number before they started playing with different combinations.

The very notion of a magic square could have come from the simplest of all number arrays:

1	2	3
4	5	6
7	8	9

Imagine trailing lines snaking through the numbers rather than the straight-line connections more familiar to us. So one might notice that the snaking line 1, 2, 5, 8, 9 adds to the same number as its complementary snaking line 3, 6, 5, 4, 7: both total twenty-five. Such trailing lines, when formalized for inscription, become more rectangular, resulting in a swastika—a form to be found in most cultures. I believe the ancients almost certainly laid the first snakes' trails across the nine integers in their most obvious order.

The ordinary listing in three columns of the first nine integers is itself, thus decoded, a magic square.

It turns out, as in figure 25.10, that the magic square revealed in the

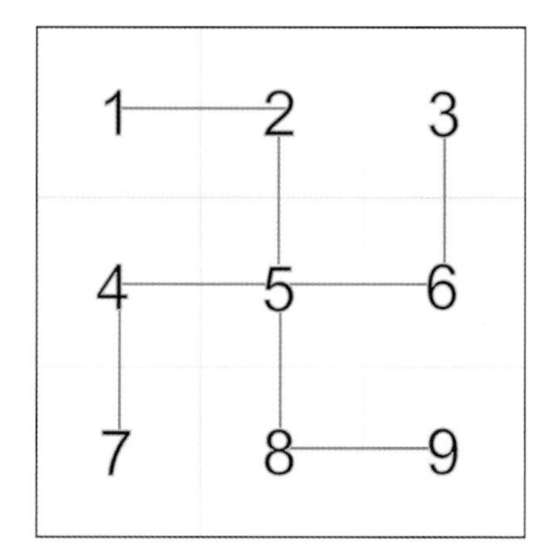

Figure 25.7
Swastika lines cover the numbers 1, 2, 5, 8, 9 and 3, 6, 5, 4, 7. Both sets add to 25—the square of 5.

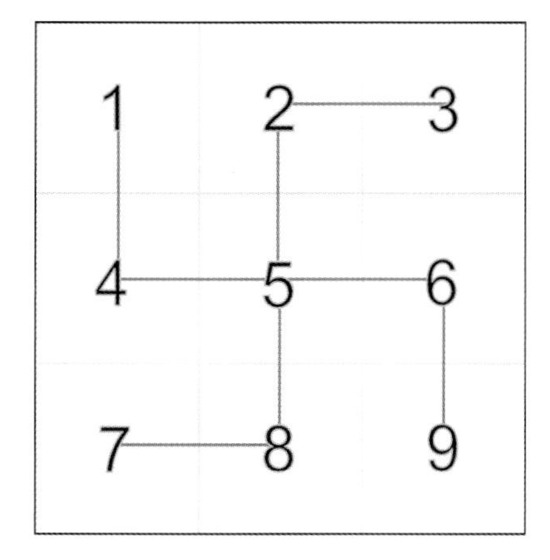

Figure 25.8
Similarly here—both sets of numbers covered by the reflected swastika add to 25.

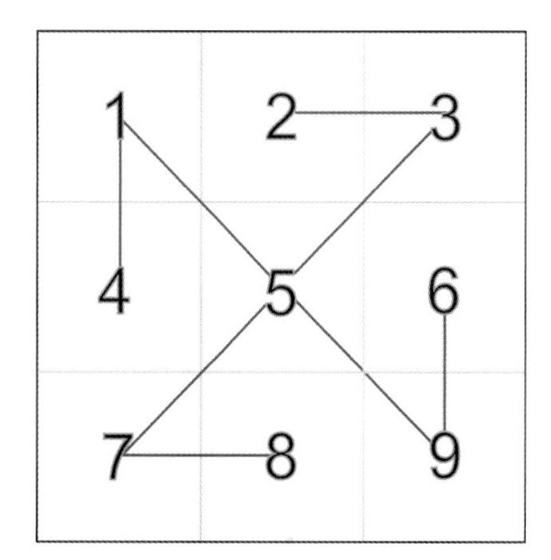

Figure 25.9
When the arrangement is on the diagonal the results are the same.

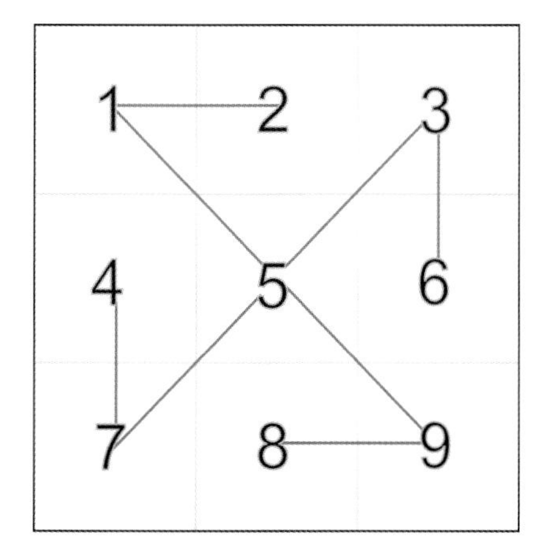

Figure 25.10
The reflected diagonal has the same total.

markings on the shell of the river turtle, despite the different number array, retains the same characteristic 25.

I can find no previous mention of this "twenty-five" characteristic of the Luo Shu. However, the concurrence of the magic square with the swastika shape was well known to scholars long before I stumbled across it. But it sprang from a different rationale.

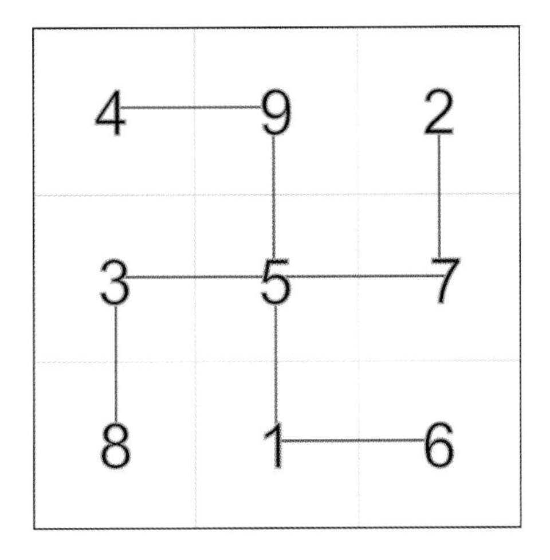

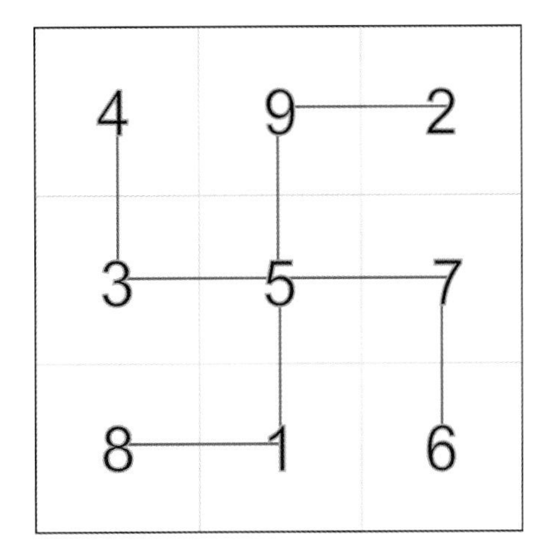

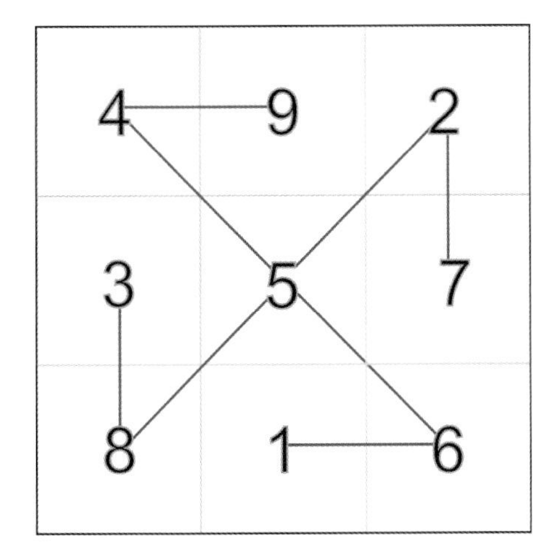

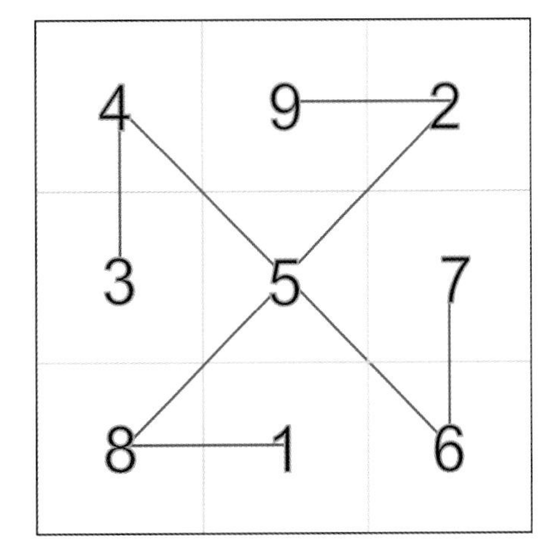

Figure 25.11

All four swastika options still add up to 25, and the numbers have now taken on their other "magic" sum quality of 15. The sum total of the numbers is 9 × 5 = 45; 5 is still physically in the middle and is the common multiple in the numerology.

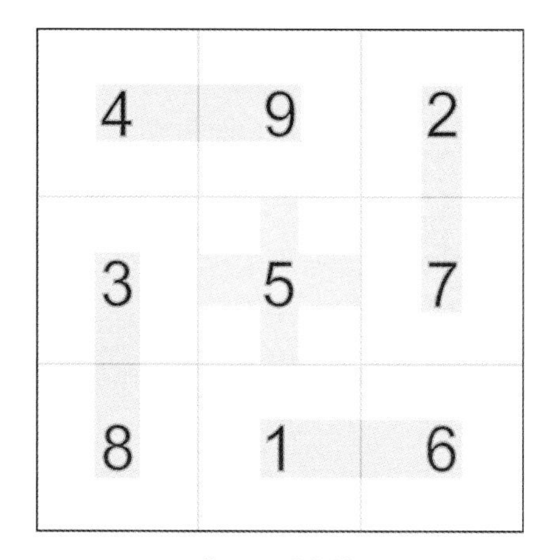

Figure 25.12
The elemental number pairs of
proto-Taoist China.

The Taoists grouped pairs of numbers with a difference of 5. A way to show this (after Berglund) is simply to write the numbers 1 to 10 in two rows:

1	2	3	4	5
6	7	8	9	10

Read vertically, this gives pairs of numbers with a difference of five between them. These pairs of numbers are those referred to earlier as the pairs denoting the elements.

> Water is the "phase" of chi that is composed by 1 (which is odd and yang) and 6 (which is even and yin).
> Fire is 2 (even yin) and 7 (odd yang).
> Metal is 4 (even yin) and 9 (odd yang).
> Wood is 3 (odd yang) and 8 (even yin).
> Earth is 5 (odd yang) and 10 (even yin).

The 5 is the difference understood as a bonding void whereby yin and yang numbers unite. Those unions are the boundary pairs with elemental ("phase") properties. Those elemental pairs interact across the—now visible and positive—5 at the center of the diagram. The result is the swastika form as generated by the philosophy of the five elements.

What has all this to do with the Starcut diagram? Numerically it was the mid-side diagonal that established the key 5 length in the Starcut. In this first layer of number we got a central 3-4-5 triangle; and √5 as half the square's side. The √5, we have also found, is the irrational constant that governs the golden number—which is also associated with our diagram via the bead numbers. These are all nice numerical correspondences but may have no historical significance. There is however something else that securely yokes this Chinese material to our study.

YU BU—THE LIMPING DANCE

For quite some years after finding what follows I believed a wrong interpretation of its historical meaning. The discovery in itself simply came from noticing the order of the numbers, from 1 to 9 in the Luo Shu. A movement traces its path through the numbers of the magic square. And the path it takes emphasizes the characteristic mid-side diagonal of the Starcut (fig. 25.13).

The movement involves two mid-side diagonals, a half side, a diagonal, another half side, and two more mid-side diagonals. It is

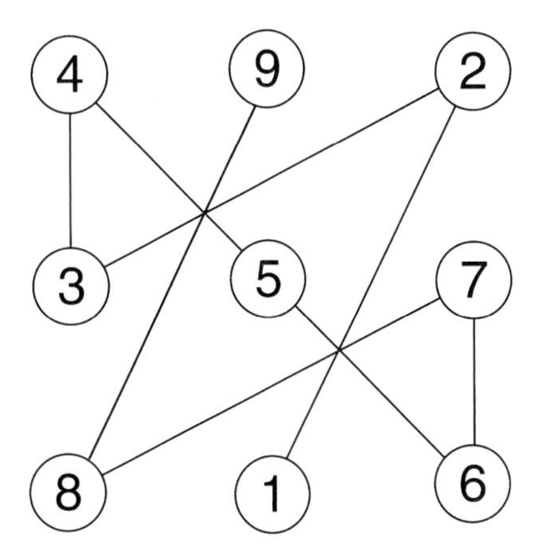

Figure 25.13
The Yu Bu movement.

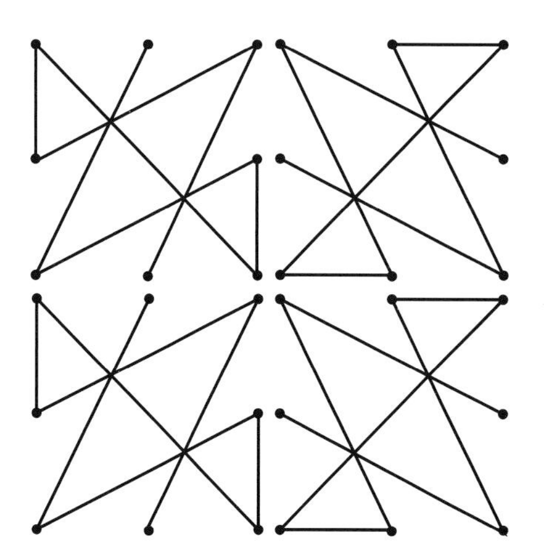

Figure 25.14
Guide patterns for variant arrangements of nine-point magic squares in the *Dao Zang*. If they are superimposed upon each other, the complete Starcut immediately appears.

almost half the Starcut; if reflected on itself it would be complete, apart from the crossbars. Rotating the numbers—with this mapping of their connections—through the different compass points seems to have been the movement of a shaman's dance. It certainly was used in the investigation of other possible magic squares as is shown in a diagram in a work called the *Dao Zang* (see fig. 25.14).

The entire Starcut is seen to be the full course of which the Luo Shu's set of steps is the basis. At first this was, to me, just a delightful coincidence. Then I read a rumor that the schema marked out the route along which the ceremonies moved through the rooms of the imperial palace (a cycle called Ming Tang). I liked the sound of this because the "route map aspect" of this magic square had already reminded me of other similar forms of movement.

The Knight's Move and "Hamiltonians"

Leaping the centuries . . . one can see in figure 25.15 how all the directions taken by pieces in the game of chess are indicated in that numbered track through the magic square. The slant of the mid-side diagonal movement is essentially that of the knight's move in chess—1 along and 2 up, or vice versa.

The precise angular slopes of knight's moves are characteristic of the relatively modern geometry of Hamiltonian circuits and courses, which are diagrams that show routes taking up all points in networks.

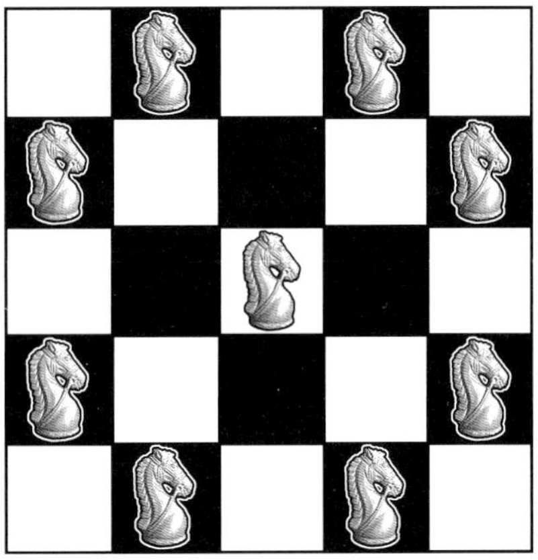

Figure 25.15
Knight's moves.

Figure 25.16
Hamiltonians in square networks. All angles correspond with those of the mid-side diagonals in the Starcut.

Hamiltonian mathematics was invented in the mid 1800s and is now a tool in genetic mapping. It relates to a similarly obscure, to me, subject: quaternions—which happened to be a great favorite with Ben Iverson, going with his four "bead" numbers. Thank heavens I am only a designer, interested simply in the fact that all the angles of slope correspond to the angles of the Starcut.

However . . .

The rumor about the ceremonial route in the imperial palace was wrong. Fortunately I showed my material to a friend Alan Hext, a master of acupuncture and a lecturer on experiential and healing aspects of Chinese culture. He introduced me to material that has finally capped my long search for a self-evident example showing the ancient provenance for this diagram.

The route through the numbers of the Luo Shu diagram was indeed a ritual movement, but not that of the emperor's Ming Tang, which had used a different schema because it needed twelve component aspects to fit with the months of the year.

This Luo Shu, on the other hand, produced a movement to do with seasonal directions and the "phases" or elements. The movement was called Yu Bu—the "Steps of Yu" also known as the "Star Walk of the Constellation of the Eight Trigrams." Because of its irregular kind of motion it was also sometimes known as "the limping step." Coincidentally this nickname applies to certain movements performed in Sufi zikrs. And one has heard that the 3-4-5 scalene triangle is, among Freemasons, sometimes called "the king of the limping triangles."

The Yu Bu was performed "shamanically" by a practitioner seen as representing, and invoking publicly, the powers of the universal Tao. The dance took in all the trigrams of the *I Ching*. It also stepped out a count of the "Nine Stars in the Heavens." Some of the visualizations indicated to be done by the practitioner, are extraordinarily vivid, and the powers invoked are accorded the status of military generals. So the dance had a martial, self-defended character infusing its stance toward the four directions.

It comes as no surprise that more recently the numerical sequence has been equated with the transfer of a practitioner's weight in the movements of the "supreme ultimate" martial art of tai chi chuan. Looking again at figure 25.13, we see that 1, 2, and 3 denote an advance to the right and a retire to the left; 4, 5, and 6 make a sweep from ahead-left to behind-lower-right, and 7, 8, and 9 depict a movement from the right across to the lower left with an advance to the center. The numbers 8, 5, and 2 have been applied to the transfer of the practitioner's chi (energy) up the left leg, through the *dan tien* balance point in the lower belly and up into the right arm and hand. I have not, thus far, found any way in which the Yu Bu steps equate directly with sequences of the foot movements in Carranza's *la destreza* rapier art, though that art was known for its dance-like steps. The coincidence of the Starcut figure in two martial forms so distant from each other both in time and space is intriguing.

The Yu Bu does, however, appear in the Western record during the Renaissance, as a glyph in Cornelius Agrippa's versions of planetary symbols, in the early 1500s (see fig. 25.17, next page). At the same time the Starcut appears in Bramante and Raphael. Presumably Agrippa realized that the graphic that he drew as the sign for Saturn was the numerical route through that planet's magic square, which he also includes, though elsewhere, in the same work.

However, soon after Agrippa's time, the seal for Saturn, while seemingly remaining associated with the magic square, loses any vestigial reference to the Yu Bu square movement. In figure 25.18, we see how the pattern becomes somewhat prettified and formalized into a circle. Would-be magicians still use this distorted sigil to signify "Saturn." Like so much ancient lore, what meaning it had has been forgotten; the dance is long lost.

In its proper form, this little dance of lines was, for the Chinese culture, universal and sacred. The number formations that it generates do not only facilitate the Luo Shu magic square; using them one can construct magic squares upon any number, and upon any set of numbers with regular intervals. In Taoist tradition it came to be used in more parochial magical invocation—precisely because of its ritual strength.

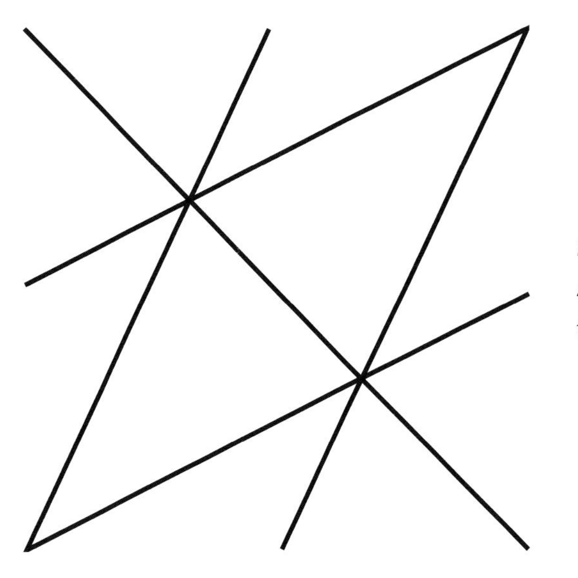

Figure 25.17
Agrippa's glyph of the Planet Saturn.

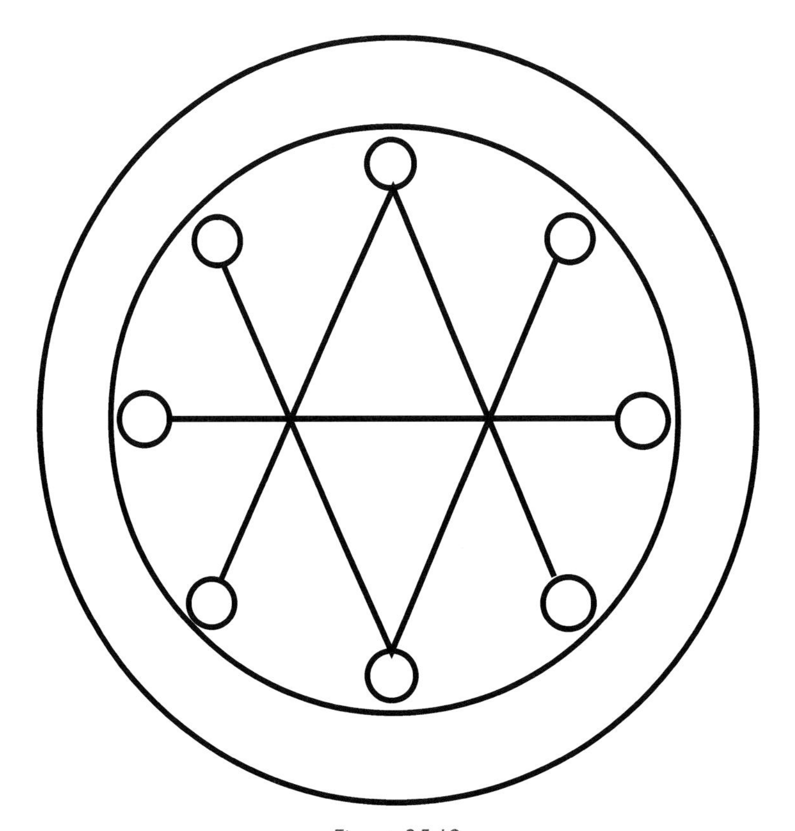

Figure 25.18
The geometric movement has been distorted out of recognition.

Berglund says of it:

> I consider the Yu Bu core itself as the great invention. The zigzag pattern acted, when used as a basic part of the yin yang context, like a genetic code, or life code. . . . There is probably no other symbol in the world which can be compared to the Luo Shu.

26
. . . Finally

I agree most heartily with Berglund's conclusion, disagreeing only when, in a passing comment, he says that he thinks it meaningless to speculate on how the Chinese came to construct the Luo Shu in the first place; as though this must have been some irrelevant fluke. There I think he misses the mark, and this whole book has, in its way, been about just that. Much that is included here seems to me to have its own intrinsic fascination and interest but within it all, I believe, there is a historical conclusion that seems to stand out.

Contemplation of the Starcut—one of the simplest, and earliest, geometric constructions—with its attendant forms and rationales has illuminated a host of cultural streams. We have seen its many connections: numerical, mathematical, musical, astronomical, geometric, martial, metaphysical, and sacred. We have found it implicit in Chinese, Vedic, Mesopotamian, Egyptian, Greek, Judaic, Sufic, Renaissance, and contemporary sources. We have gained thereby what is, at the very least, an extraordinary mnemonic for a huge swath of the perennial arts. I have referred to these arts as "Apollonian" because, in addition to their practical applications, they seem always to have been intimately associated with the soul. However, I believe it plausible that this device is more than a modern mnemonic projected upon the past.

Opposite: Figure 26.1 The Starcut diagram as a tessellated mosaic.

Demonstrating the plausibility of something, of course, does not prove that thing to be true. Plausibility is both a blessing and a curse.

Without it we would not have noticed a vast range of now-proven propositions that underlie all that is sure in our sciences. It can, however, also be a tool of deception. Many of those who influence our political, commercial, journalistic, social, and intellectual culture use plausible associations to manipulate agreement, to suggest unreliable courses of action, and to propose opinion as truth.

I cannot prove what I suspect about the Starcut diagram. I suspect that it was an immensely early graphic device predating Euclid by thousands of years. It is not something that grows from sophisticated conclusions about geometry; it is very simple, though as we have seen its applications are vast. I take it to have been a conceptual foundation for the edifice of explorations, recognitions, and practical constructions that, with the distillation of many centuries, emerged as Euclid's *Elements*.

Where did the very logic that Euclid used have its basis? It too needed its own self-recognition and formulation. It was Aristotle who did that in identifying the laws and the twenty-seven core syllogisms of formal logic; but he did not invent the pattern of reasoning that he defined. In Aristotle the reasoning faculty applied itself to itself in order to define its own boundaries. Plato had already demonstrated the *application* of such logic in the dialogues of Socrates. Plato in turn knew that fundamental right reasoning was involved in the "geometry" that he required of all who even approached the threshold of his Academy. And we can trace the threads of geometry, number, and the attendant logic they demand (even if it was not yet self-conscious) right back beyond the historical record. The great mountain of Mesopotamian cuneiform clay stelae, only a small proportion of which have been translated and studied, prove that the analytical logic of causes, calculation, and consequences was alive and well in deepest antiquity; and the rationales detectable in the calendrical structures of even earlier megalithic cultures show exactly the same thing, as do the ground plans visible in the ruins of Jericho, Çatal Hüyük, and Mohenjo Daro—all more than seven thousand years old.

The astronomer Fred Hoyle writing about Stonehenge made just this point:

> The intellectual activity of mankind during prehistory is a vast almost uncharted ocean. . . . There have been only about 200 generations of history and upward of 10,000 generations of prehistory.

He also observed that the studies of those generations would have been done "with a basic intelligence equal to our own."

Because of the way we have learned about geometry and geometric numeration Euclid and his near-in-time Hellenic predecessors seem to have been originators, but at the very horizon where prehistory begins to become history there are deep foundations of practical and theoretical knowledge already in place.

The recognition of the right angle, of similar triangles of the same shape but of different size, the concepts of division and fractional ratios, the measurement of areas and the equivalence of areas between different shapes, the self-consistent sexagesimal number system, the use of

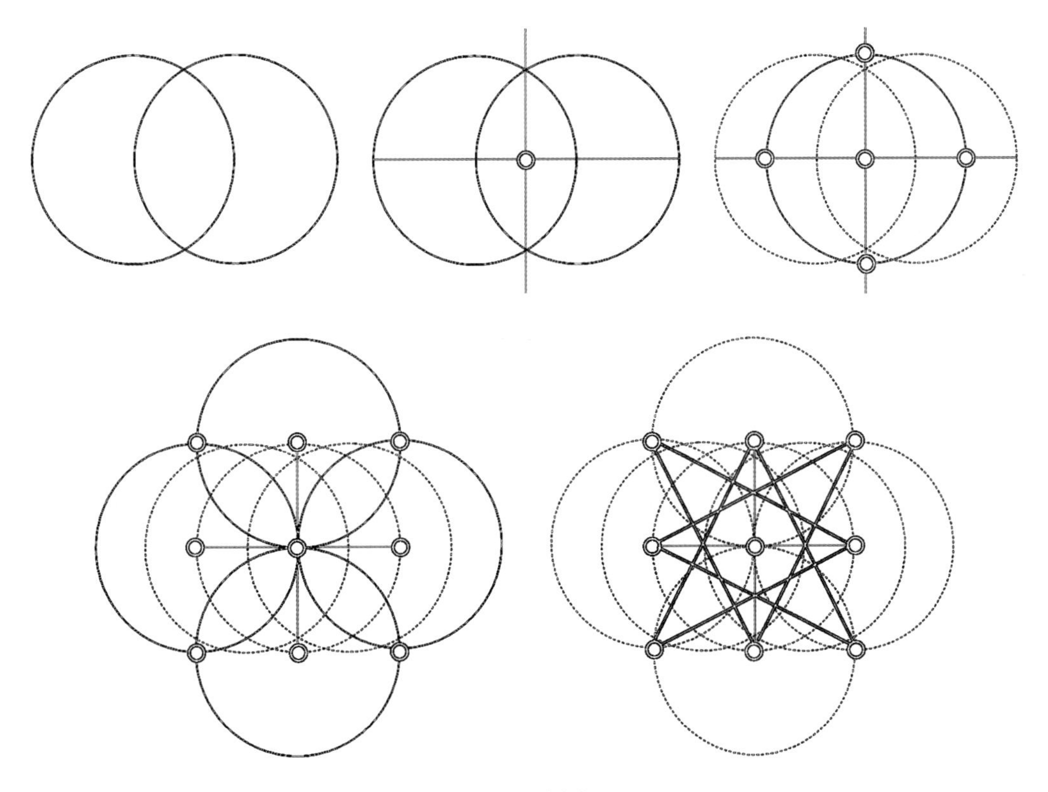

Figure 26.2
The seven circles in the sand.

geometry to drive forward the understanding of number and the special relationships of certain number sets—all these and much more became part of human lore at a time that remains out of sight. Physical constructions embodying such recognitions cast light a little further back, and by (highly plausible) deduction they imply the seed forms of geometry itself. They were known through the stretching of cords and the connection of points. They brought us from the single circle to the seven circles and the nine points that gave us a primary key by which we learned to see, hear, enumerate, and integrate the mind with patterns of perceived harmony, form, and number.

So much for the historical aspect of this book. The other main aspect of it has been the geometric/numerical essence of the Starcut diagram. Here I would like to hazard a theoretical suggestion as to how it comes to be so productive of geometric insight.

The "journey" of geometry from a triangle to a pentagon has to cross the square. The triangle, as we saw at the very beginning, grows out of the twofold vesica. It proliferates in equilateral form through clusters of circles. In a square, the most basic triangle that can be made is by a diagonal. Diagonals can be used to cut up squares endlessly in twofold multiplication. The geometry of the square's ordinary diagonal leaves us simply with the square itself in smaller and smaller form (fig. 26.3).

As noted earlier, what the Starcut does is to bring 5 into the game. The mid-side diagonal is the hypotenuse across the span of two lines in 1:2 ratio. We see this in figure 26.4.

This line immediately introduces 5 and in more than one way. The hypotenuse of that triangle is immediately $\sqrt{5}$, in terms of the 1:2 ratio of its sides. If a second mid-side diagonal is drawn so as to intersect with the first, the lines are found to have divided each other into fivefold sections. Those sections, as we have seen many times throughout this study, provide a whole system of numbers wherein the mid-side diagonals become of length 30 (6 × 5), and the total square area becomes 720 (144 × 5). Fivefold sections, multiples, and square roots now proliferate, and it is the sexagesimal number system that characterizes the triangular areas of the whole field. The same system, and particular key numbers within it, we found to concur with the numerical Tetraktys.

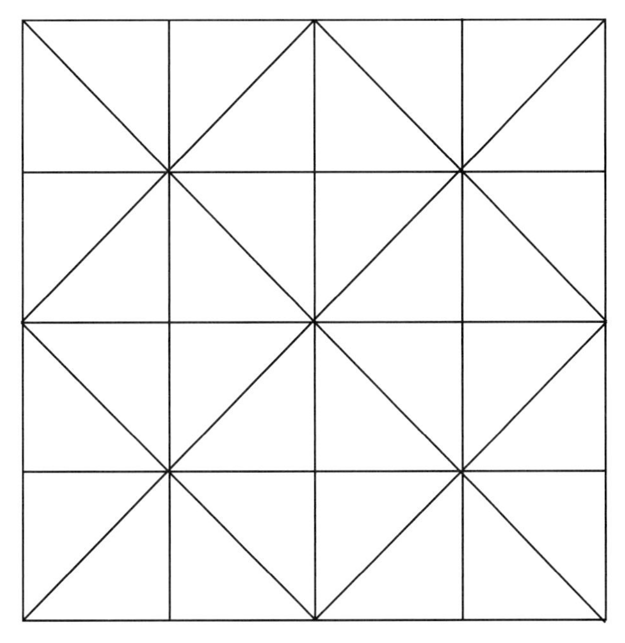

Figure 26.3
Diagonal cuts in the square.

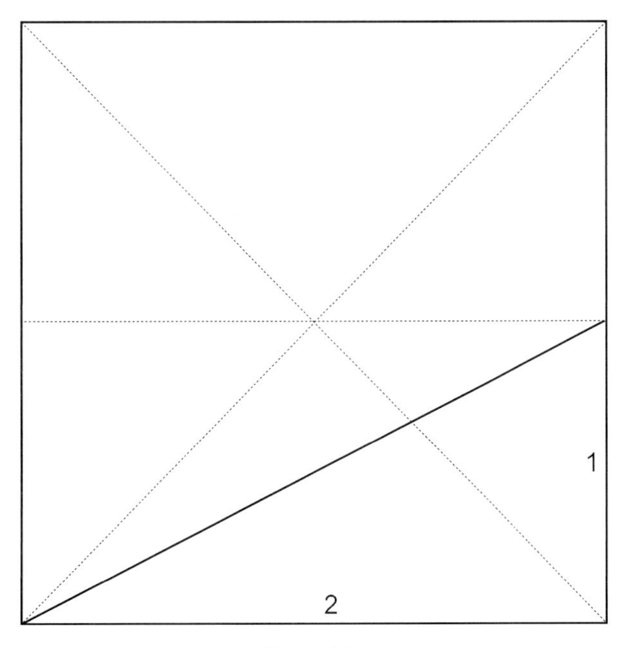

Figure 26.4
The mid-side diagonal.

What is notable about all this is that the construction that generates it is not some imported contrived pattern. It is simply the result of using the information latent in drawing the square in the first place.

Straightaway that one does extend the construction by using the mid-side diagonal as a radius to swing an arc, outside the square, one thereby produces the geometry of ϕ and the key to pentagon construction.

So the mid-side connections have "laid the square open." Its inward structures, its musical proportions and geometric rationales are revealed. The same lines also release the square's external potentials way beyond the repetitive geometry based on the diagonal alone. They show the fivefold possibilities of four-ness; nor do they stop there, because the nodes where these same lines intersect result in the infinite rational divisibility of the square's sides and areas. And it all comes from a shape so simple that it can be drawn with a forked stick in the sand.

Looked at through the Starcut lens, sacred geometry becomes one thing.

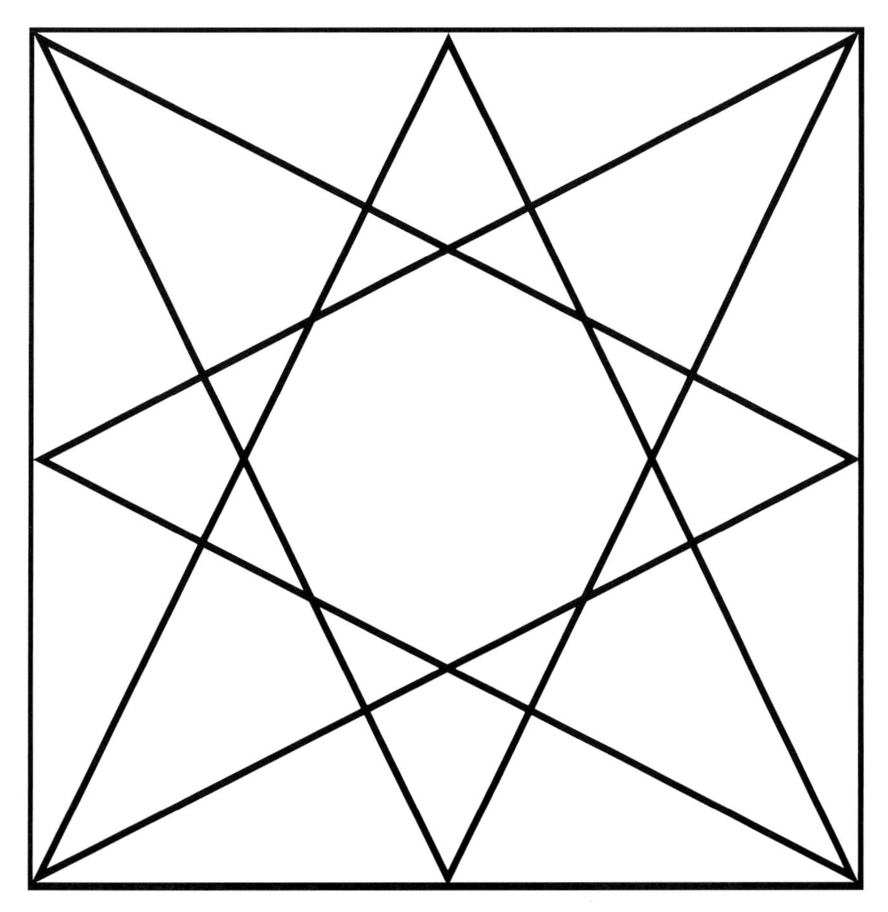

Figure 26.5
The Starcut.

Epilogue

The sun had set and the twilight caught the crests of waves on the incoming tide.

"I'm a bit numb," said the southerner.

"Me, too," said the monk.

A seething tongue of foam lapped up the beach, and the shadowy edges of the sand diagram were engulfed. The water receded, leaving just the merest trace as the men turned to go.

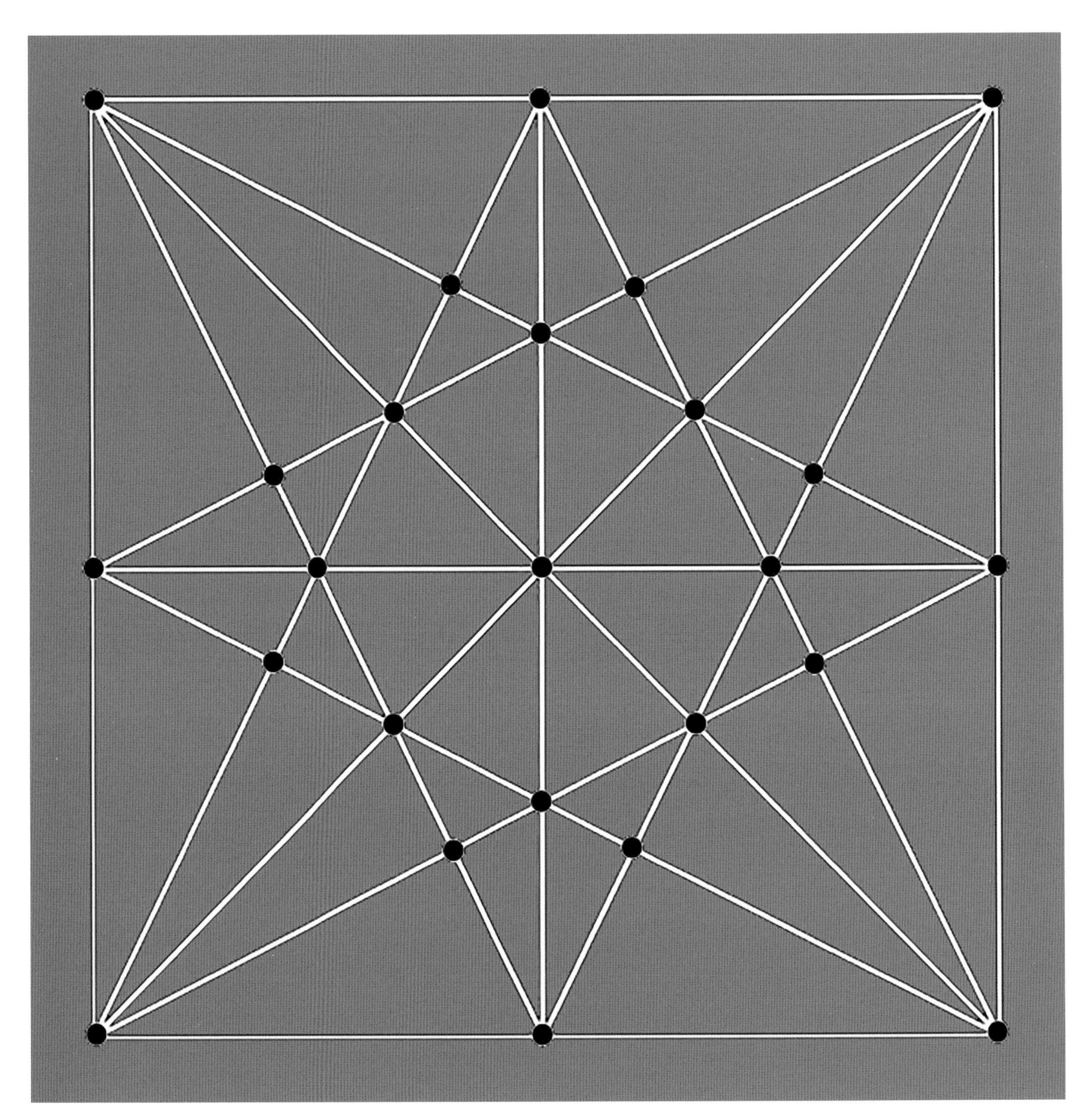

The game board with its nodes marked.

APPENDIX

The Starcut Glass Bead Games

Conceived and designed by Malcolm Stewart

THE GAME BOARD

The geometric matrix—the game board—is a square with its central meridians and its diagonals and with eight further lines, two from each corner to the midpoints of their opposite sides.

All the glass bead games are played by placing and/or moving beads on the nodes where the lines of the lattice cross or connect. There are twenty-five nodes. Directions for each of the five games are given on the pages that follow.

THE GAMES

Glass Bead Game 1

Yubu—The "Third Bead" Game

- Two players each have twelve beads of different color.
- Players play single beads alternately, placing them on the nodes of the lattice.
- Once played, a bead cannot be moved (unless removed—see bullet 7).

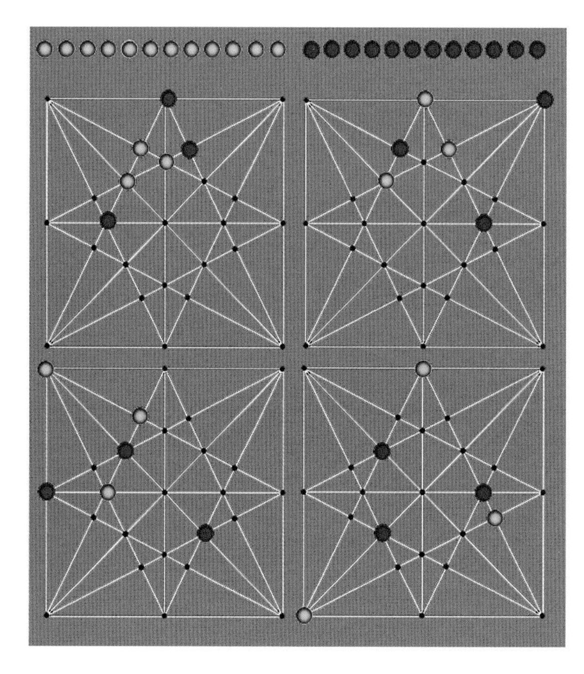

Game 1.1
The green beads show four different sizes of 3-4-5 triangles and some of the thirty-two winning combinations.

- The first player who occupies the corner nodes of a 3-4-5 triangle wins the game. There are thirty-two such triangles in four different sizes. The green beads in the examples on the right illustrate the four different sizes of such triangles and some winning combinations (and see next pages). In actual play, it is improbable that someone would win after playing only three beads.
- Players can decide whether a player must say "Check" when placing a bead that threatens a win on their next move.
- A player cannot play a third bead on a line that already has two of their own beads on it, unless that third bead either wins the game or escapes from check.
- If a player plays such a third bead, the other player says "Third bead" and removes it and continues with their own play with the advantage of an extra bead. If, however, a "third bead" challenge is made incorrectly (see below), the challenger forfeits their next play.
- If nodes on all available lines are already occupied by two of a player's beads, that player must play on a line that has the least of their own beads on it. It is in this case that there may be an incorrect challenge.

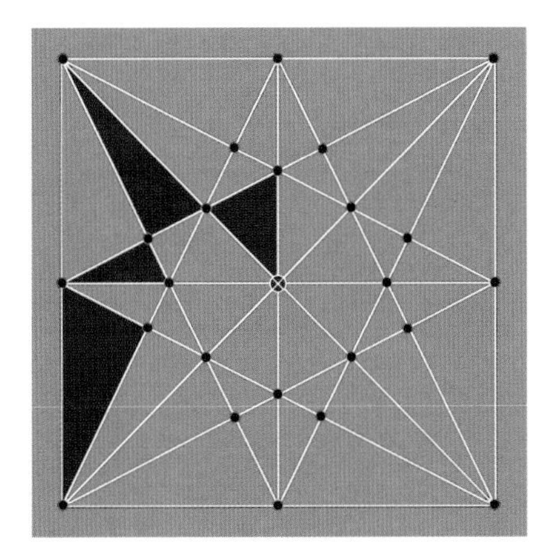

Game 1.2

In dark gray here are examples of triangle shapes that do not win. In the Yubu Game the central node is never part of a winning shape, and so remains unoccupied.

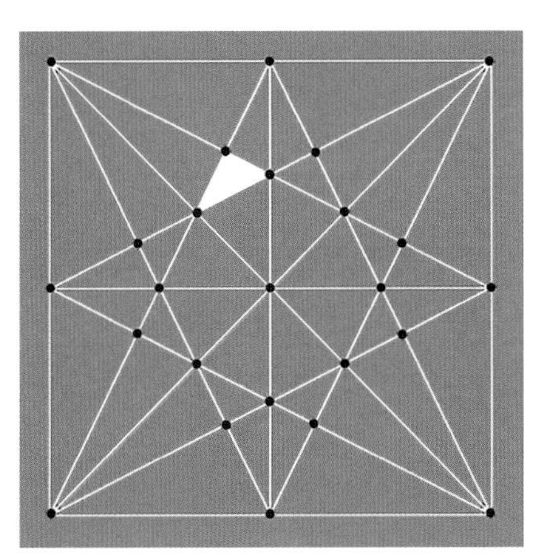

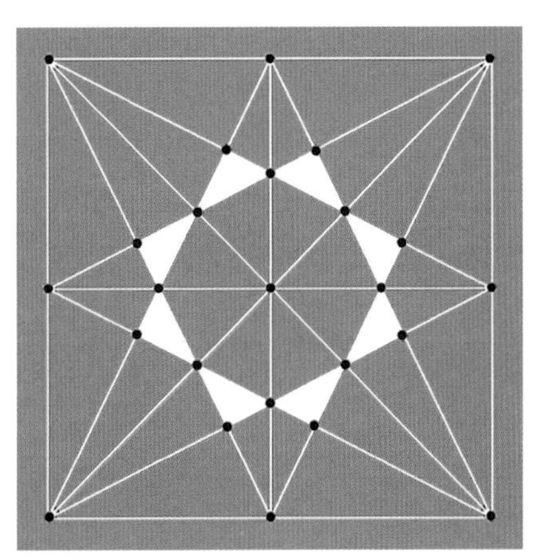

Game 1.3

Below is a 3:4:5 proportion triangle. There are eight triangles of this form and size. A player occupying the three corner nodes of any one such triangle wins the game.

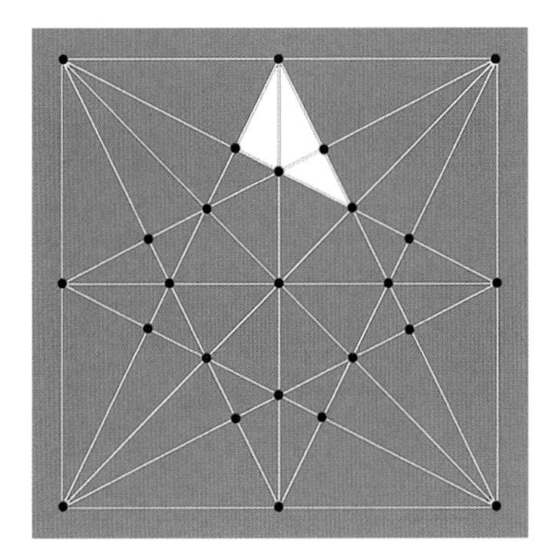

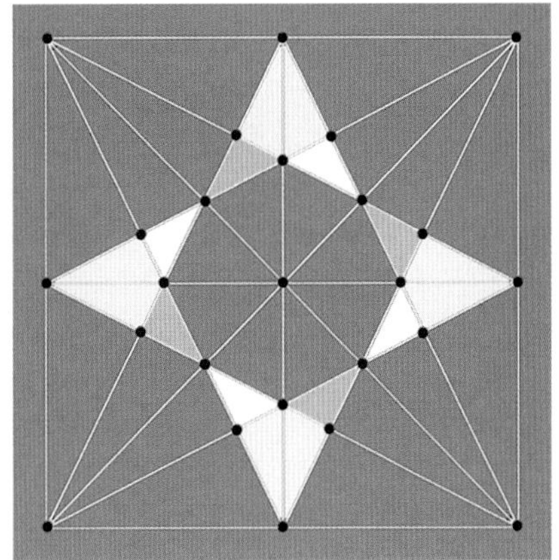

Game 1.4

These are the eight next-larger versions of the same shape triangle. Note how they overlap. Two such triangles are found at each mid-side node. Again, a player wins who occupies the corner nodes of any of these triangles.

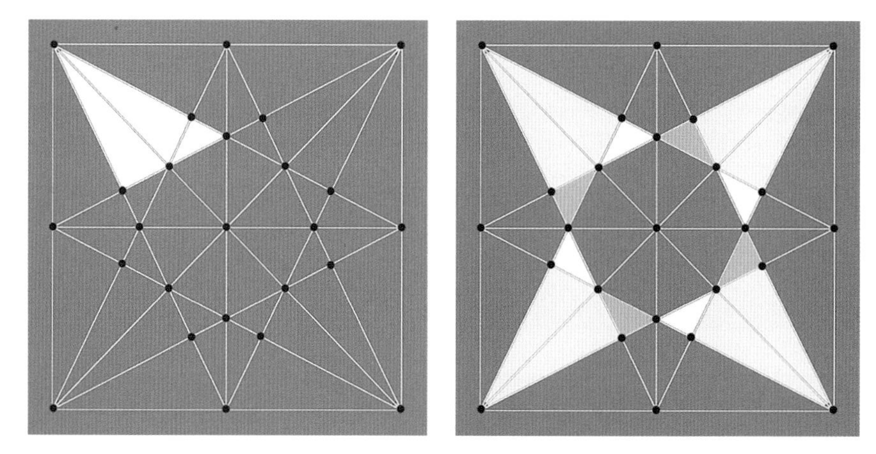

Game 1.5

Here are triangles of the same shape in the largest size. Two such are found at each mid-side. Again, a player who occupies the corner nodes of any of these triangles wins.

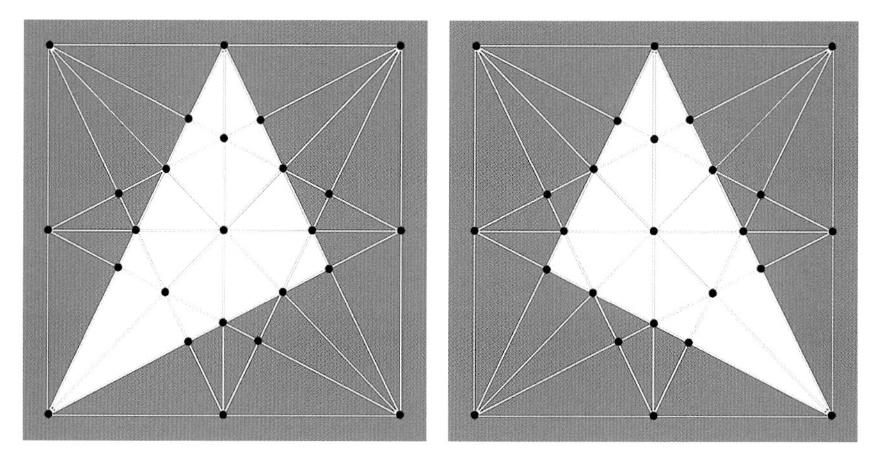

Game 1.6

The first player to occupy the corners of any one of the thirty-two triangles of this proportional shape wins, irrespective of the size of the triangle.

Glass Bead Game 2

Starwalk—The "Fourth Bead" Game

- Two players each have seven beads in two different colors.
- In the "placing phase" of the game, players place single beads in turn, putting them on any node of the lattice, until all seven are placed.
- Thereafter the "moving phase" commences, in which players take turns to move beads from node to node along lines.
- In one turn a player can move any one bead along any line on which it is situated (all nodes are on at least two lines); a player's move may be across as many nodes as are unoccupied by some other bead, whether their own or their opponent's. Note that players can move along the edge of the lattice.
- Within a single turn players cannot move around corners but can only go straight along lines. They cannot jump their own or the other player's beads.

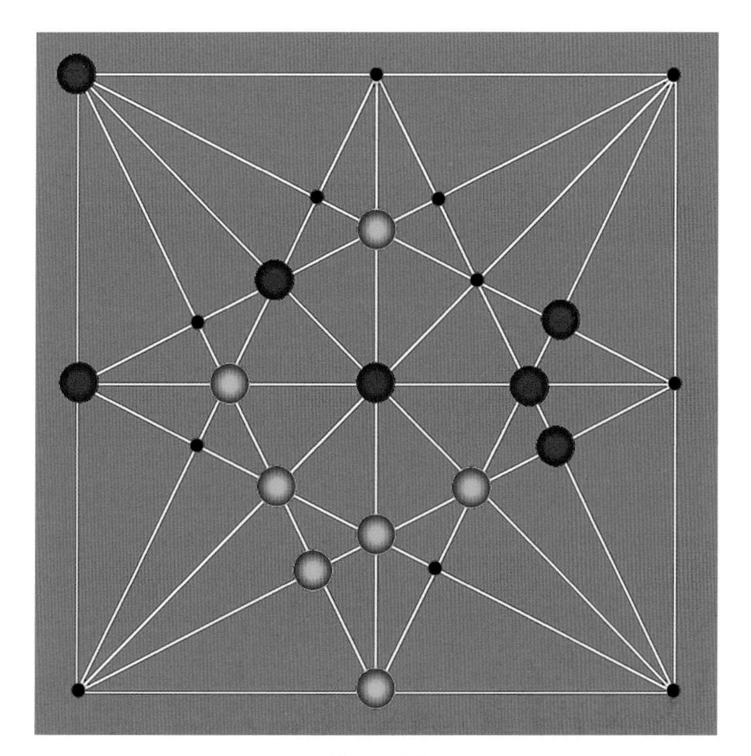

Game 2.1

In this example green wins; red having previously had to move to the central node to avoid defeat.

- The aim is to make four beads on adjacent nodes along any line. The player to do this wins. The game may end before the "moving phase" since it is possible that a player may make four in a row at an earlier stage.
- If both players achieve four in a row within the same number of moves, the game is a draw unless either player can, immediately thereafter, place a fifth bead in the same row as their four. That player then wins.
- A bead can only be placed on, or moved to, the central node if it thereby avoids defeat on the opponent's next turn, or, by being so placed or moved, itself becomes the fourth bead in a winning combination. It cannot be so placed or moved to avoid the opponent getting three beads in a row.

Glass Bead Game 3

Boxing Bead Game

- Two players each have seven beads of different color.
- Players take turns placing, and later moving, their own beads on the nodes of the board with the aim of winning by "boxing" a square through occupying its four corner nodes.
- The placing of each player's seven beads must be complete before either player can start to move their beads. The game may be won during this first stage if a square is boxed.
- In the moving stage: players take turns to move single beads. A bead can only be moved along some line that passes through the node where the bead in question is situated, and the bead can only move in one direction during one turn. It can be moved more than one node point, however, if the player so wishes and the way is clear. A move must stop on the node adjacent to one that is occupied by another bead. A bead cannot jump over any other bead.
- At any stage of the game players can box in their opponent's beads and remove them from the board. If a bead cannot move (because there is no free point adjacent along a line from their position) it is "boxed." It is immediately removed. The pieces boxing in the lost

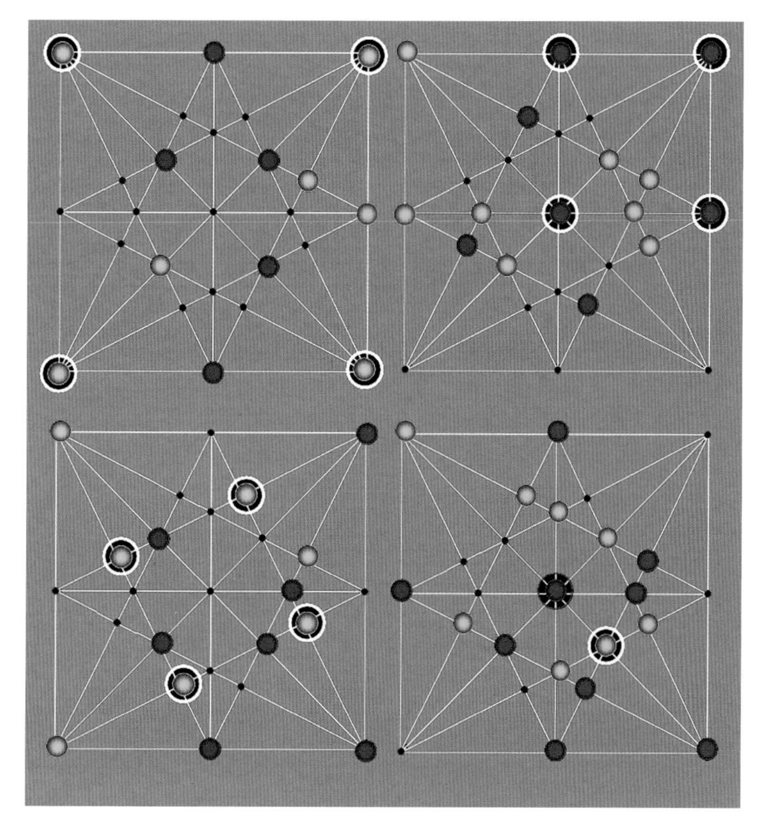

Game 3.1

Sample boxed squares—there are seven possible squares.

Top left: the main square—here green wins.

Top right: four quadrant squares—here red wins.

Bottom left: there are two tilted central squares—here green wins.

Bottom right: the ringed bead cannot move and is taken off.

bead may belong to either player, as in the example on the extreme bottom right.

- The central node can only be used for "taking" (occupying the last possible move for one of the opponent's beads), for a winning move (completing a box), or as an escape from loss either of a bead or of the game. See example on top right of the adjacent figure; the ringed red bead will have been the last play to take the green.

Glass Bead Game 4

Rapier

- Two players each have six, seven, eight, nine, ten, or twelve beads of different color.
- Players may array their beads as they wish, each in one-half of the lattice. At the outset, for games with ten or fewer beads per player, there must be a clear half-lattice dividing line between the two players' beads—either a meridian or a diagonal—with its five nodes

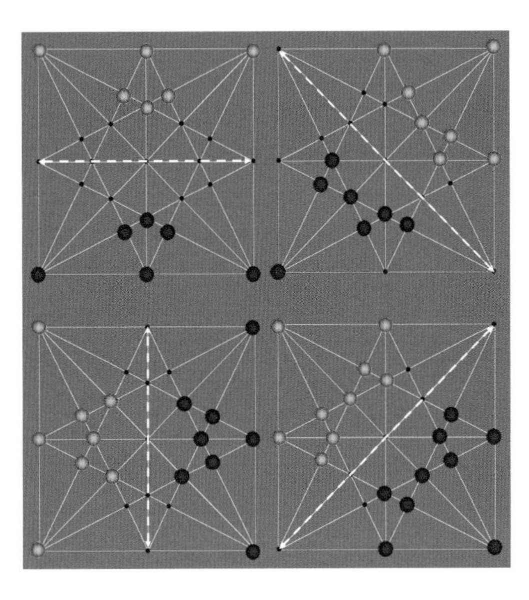

Game 4.1

Sample opening arrays shown above and below the dotted white lines to mark agreed half-board dividing lines.

Symmetrical 6 versus 6 array;

Asymmetrical 7 versus 7 array;

Symmetrical 8 versus 8 array;

Asymmetrical 9 versus 9 array.

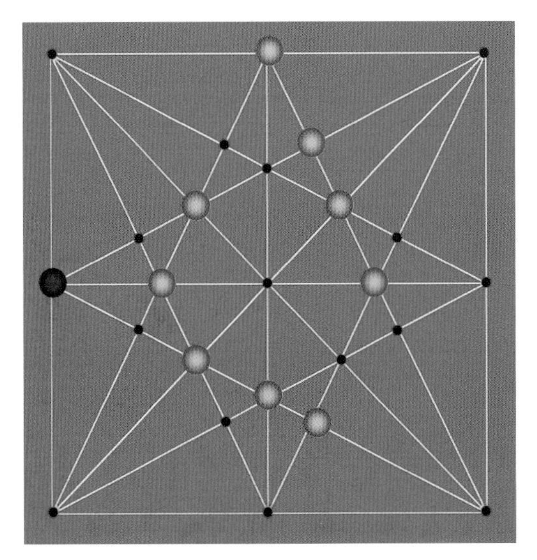

Game 4.2

Here is a jumping bead game problem that illustrates extreme possibilities within a single turn. It is a nine-bead game and red's turn. Red is to play and win!

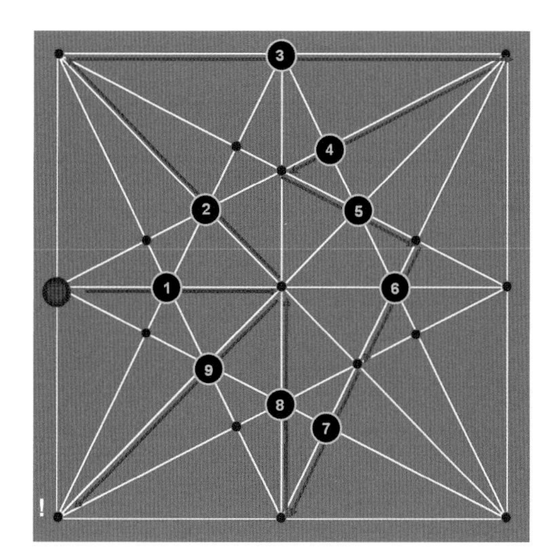

Game 4.3
There are a number of solutions where red does damage but loses. A nine-bead solution is indicated by the ringed numbers showing the order in which the green beads are jumped and removed. The shown red bead is the starting node and the exclamation point (!) marks red's final, and winning, position.

unoccupied. (See samples in game 4.1. For the twelve-bead game, see below.)

- This dividing line may be occupied or crossed once play commences.
- Players take turns to move, and move one bead at each turn.
- All moves must be jumps either over one's own or one's opponent's beads, or over a combination of both players' beads.
- A jump can only be made if the starting node is adjacent to a bead that has a free node on its other side along a line, as in Chinese checkers. A player can keep jumping in any direction and over as many beads as are situated so as to allow jumps. All lines of the lattice may be used, including the edges and diagonals,
- As a player jumps an opposing bead, that bead is removed from the board. A player's own beads are not removed when jumped. It is important to remove opponent's beads immediately as they are jumped, because this sometimes clears the way for later jumping possibilities within the same turn.
- A player wins when the opponent has lost all their beads or can no longer move the bead or beads remaining to them (i.e., can no longer make any jump). To win, however, a player has to be able to make at least one further jump—otherwise the game is a draw.

Twelve-Bead Rapier

The player whose beads occupy five of the outer edge points (red in fig. 4.4/5.1), takes the first turn.

The twelve-bead game has to start with a take.

Glass Bead Game 5

Sand Reckoner Solitaire

- This is similar to traditional solitaire but on the glass bead game lattice.
- It is a game for a single player.
- Twenty-four beads of whatever color are arrayed so that only the central node is vacant.
- All moves must be jumps over adjacent beads to land on open nodes. All jumped beads are removed from the board. The game's objective is to end with only one bead remaining, and that bead is to be situated on the central node.

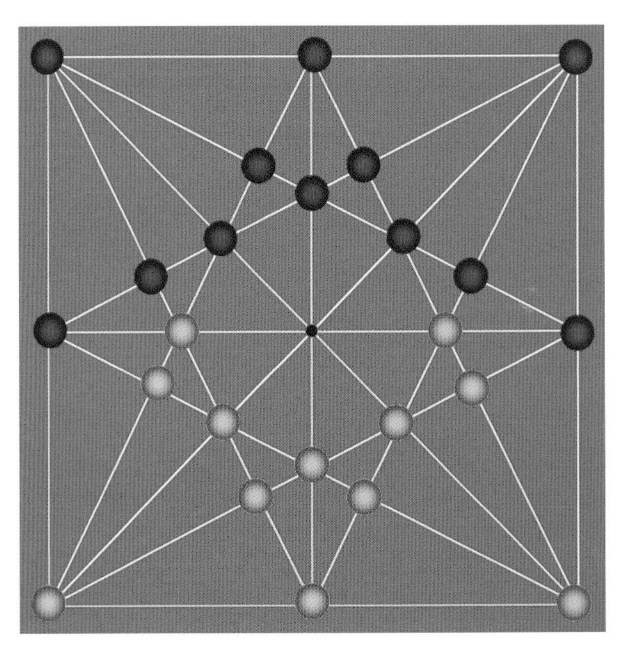

Game 4.4/Game 5.1

The twelve-bead layout with red to start.

This is also the starting array for Starcut Solitaire (game 5).

Bibliography

Albarn, Keith, Jenny Miall Smith, Stanford Steele, and Dinah Walker. *The Language of Pattern*. London: Thames and Hudson, 1974.

Ardalan, Nader, and Laleh Bakhtiar. *The Sense of Unity: Ths Sufi Tradition in Persian Architecture.* Chicago: University of Chicago Press, 1973.

Ball, Philip. *The Self-Made Tapestry: Pattern Formation in Nature.* Oxford: Oxford University Press, 1999.

Bond, Frederick Bligh, and Lea Simcox. *The Apostolic Gnosis.* London: Research Into Lost Knowledge Organisation, 1979.

———. *Gematria.* London: Research Into Lost Knowledge Organisation, 1977.

Bourgoin, J. *Arabic Geometrical Pattern and Design.* New York: Dover Publications, 1973.

Brunes, Tons. *The Secrets of Ancient Geometry.* 2 vols. Copenhagen, Denmark: Rhodos, 1967.

Charpentier, Louis. *The Mysteries of Chartres Cathedral.* London: Research Into Lost Knowledge Organisation, 1966.

Cole, J. H. "Determination of the Exact Size and Orientation of the Great Pyramid of Giza." Survey of Egypt paper no. 39. Cairo: Government press (1925).

Cowan, Painton. *Rose Windows.* London: Thames and Hudson, 1979.

Cox, Kathleen. *Vastu Living.* New York: Marlowe and Company, 2000.

Critchlow, Keith. *Order in Space.* London: Thames and Hudson, 1969.

———. *Time Stands Still.* London: Gordon Fraser, 1979; reissued Edinburgh: Floris Books, 2007.

Critchlow, Keith, Robert Lawlor, Christopher Bamford, Arthur Zajonc, Anne Macaulay, and Kathleen Raine. *Homage to Pythagoras: Rediscovering Sacred Science*. Great Barrington, Mass.: Lindisfarne Books, 1994.

Daraul, Arkon. *Secret Societies*. London: Octagon Press, 1961.

Fauvel, John, and Jeremy Gray. *The History of Mathematics: A Reader*. London: Macmillan with Open University Press, 1987.

Fideler, David R., ed. *The Pythagorean Sourcebook*. Grand Rapids, Mich.: Phanes Press, 1987.

Fletcher, Sir Banister. *A History of Architecture*. London: Athlone Press, University of London, 1961.

Furlong, David. *The Keys to the Temple*. London: Piatkus, 1997.

Ghyka, Matila. *The Geometry of Art and Life*. New York: Dover, 1977.

Godwin, Joscelyn. *The Pagan Dream of the Renaissance*. London: Thames and Hudson, 2002.

Guthrie, Kenneth Sylvan. *Pythagorean Sourcebook and Library*. Grand Rapids, Mich.: Phanes Press, 1987.

Hall, Manly P. *The Secret Teachings of All Ages*. Los Angeles: Philosophical Research Society, 1977.

Hancox, Joy. *The Byrom Collection*. London: Jonathan Cape, 1992.

Haynes, Ofmil C. *The Harmony of the Spheres*. Powys, Wales: Wooden Books, 1997.

Heath, Richard. *The Matrix of Creation*. Cardigan, Wales: Bluestone Press, 2002.

Heath, Robin, and John Michell. *The Measure of Albion*. Cardigan, Wales: Bluestone Press, 2004.

Iamblichus. *The Theology of Arithmetic*. Translated by Robin Waterfield. Grand Rapids, Mich.: Phanes Press, 1988.

Iverson, Ben. *Pythagoras and the Quantum World*. New York: Carlton Press, 1982.

Jeans, Sir James. *Science and Music*. New York: Dover Publications, 1968.

Johnson, Anthony. *Solving Stonehenge*. London: Thames and Hudson, 2008.

Jones, Sir William. *Asiatic Researches: Transactions of the Society Instituted in Bengal*. Vol. 2, *On the Chronology of the Hindus*. Calcutta: Robert le Blond, 1798.

Khan, Hazrat Inayat. *The Sufi Message of Hazrat Inayat Khan*. Vol. 2, *The Mysticism of Sound*. Southampton: Camelot, 1962.

Lahanas, Michael. "Archimedes and Combinatorics (The Loculus of Archimedes)." Hellenica World website.

Lawlor, Robert. *Sacred Geometry.* London: Thames and Hudson, 1982.

Lefebure, F. *Phosphenism: The Art of Visualisation Developing Memory and Intelligence.* Fleet, Hants: Psychotechnic Publications, 1990.

Lilly, John C. *The Center of the Cylone.* New York: Julian Press, 1972.

Martineau, John. *Altair Raindrops.* Powys, Wales: Wooden Books, 2000.

———. *A Little Book of Coincidence.* Powys, Wales: Wooden Books, 2001.

McClain, Ernest G. *The Myth of Invariance.* Boulder, Colo.: Shambhala, 1978.

Michell, John. *The Dimensions of Paradise.* London: Thames and Hudson, 1988.

———. *The New View over Atlantis.* London: Thames and Hudson, 1983.

Stirling, William. *The Canon.* London: Garnstone Press, 1974.

Taylor, Thomas. *The Theoretic Arithmetic of the Pythagoreans.* London: Prometheus Trust, 2006.

Thibault, Gerard, *Academy of the Sword.* Translated by J. M. Greer. Highland Village, Tex.: Chivalry Bookshelf, 2006.

Yates, Frances A. *The Art of Memory.* London: Pimlico, 1992.

Index

Page numbers in *italics* indicate illustrations.